Beach Management

Coastal Morphology and Research

Series Editor: Eric C.F. Bird

CORAL REEF GEOMORPHOLOGY

André Guilcher

COASTAL DUNES
Form and Process

Edited by
Karl Nordstrom, Norbert Psuty and Bill Carter

GEOMORPHOLOGY OF ROCKY COASTS

Tsuguo Sunamura

SUBMERGING COASTS
The Effects of a Rising Sea Level on Coastal Environments

Eric C.F. Bird

BEACH MANAGEMENT

Eric C.F. Bird

Beach Management

Eric C. F. Bird
Geostudies

JOHN WILEY & SONS
Chichester • New York • Brisbane • Toronto • Singapore

Published 1996 by John Wiley & Sons Ltd,
Baffins Lane, Chichester,
West Sussex PO19 1UD, England

National 01243 779777
International (+44) 1243 779777

e-mail (for orders and customer service enquiries): cs-books@wiley.co.uk.
Visit our Home Page on http://www.wiley.co.uk
or http://www.wiley.com

Other Wiley Editorial Offices

John Wiley & Sons, Inc., 605 Third Avenue,
New York, NY 10158-0012, USA

Jacaranda Wiley Ltd, 33 Park Road, Milton,
Queensland 4064, Australia

John Wiley & Sons (Canada) Ltd, 22 Worcester Road,
Rexdale, Ontario M9W 1L1, Canada

John Wiley & Sons (Asia) Pte Ltd, 2 Clementi Loop #02-01,
Jin Xing Distripark, Singapore 0512

Library of Congress Cataloging-in-Publication Data

Bird, E.C.F. (Eric Charles Frederick), 1930–
Beach management / Eric C.F. Bird.
p. cm. — (Coastal morphology and research)
Includes bibliographical references (p. –) and index.
ISBN 0-471-96337-2
1. Shore protection. 2. Coastal zone management. I. Title.
II. Series.
TC330.B57 1996
627'.58—dc20 96-11165
CIP

British Library Cataloguing in Publication Data

A catalogue record for this book is available from the British Library

ISBN 0-471-96337-2

Typeset in 10/12pt Times from author's disks by
Mayhew Typesetting, Rhayader, Powys
Printed and bound in Great Britain by
Biddles Ltd, Guildford, Surrey

This book is printed on acid-free paper responsibly manufactured from sustainable forestation, for which at least two trees are planted for each one used for paper production.

Contents

Preface

This book is an account of the origin, evolution and changing features of beaches, prepared as a reference work for people concerned with coastal planning, management, engineering and development. It is based on my experience as a geomorphologist advising on coastal problems in the course of work with the Port Phillip Authority and other coastal management agencies in Australia, and projects carried out on behalf of the United Nations University and the United Nations Environment Programme in south-east Asia, the Americas, Britain, Europe and the former Soviet Union.

Beaches have attracted people to the coast for more than two centuries, and large numbers visit beaches at seaside resorts and tourist areas each year for holidays and recreation. Beach utilisation has been of increasing importance to the economies of coastal regions, and beach management has become necessary to deal with a variety of problems. There has, for example, been much concern over problems of beach erosion, which was shown in a previous book in this series, *Coastline Changes: A Global Review* (published in 1985), to have become prevalent around the world's coastlines during the past few decades. Beaches are a diminishing resource, and the space available for beach recreation has been declining. This book includes a review of the causes of beach erosion, and goes on to consider various responses made to it, including the building of shore protection structures, some of which have had adverse effects. In recent years beach nourishment has become widely used as a means of maintaining or improving a beach, and the principles that should guide beach nourishment are discussed, along with some of the problems that have arisen, and the ways in which beaches can be stabilised. Beach management will be most effective if those concerned understand how their beach has taken shape, what changes are occurring and why, and what is likely to happen to it in the future. Beach sediments are resources that need to be conserved and supplemented if beach environments and opportunities for beach recreation are to be maintained.

Prediction of beach changes must now also take account of the probability that the sea level will rise over the next few decades, with consequences that were discussed in another book in this series, *Submerging Coasts: The Effects of a Rising Sea Level on Coastal Environments* (published in 1993). However, the immediate task is to deal with changes already taking place and the impacts of various activities and structures on beaches.

Beaches are accumulations of sediment – mainly sand and pebbles – and this book is concerned particularly with the origins and movements of beach sediment that result in gains and losses, both natural and artificial, on the world's coastlines. Nearshore processes – waves, tides, currents and wind action – are considered in sufficient detail to explain the changes taking place. If further information on marine process systems is sought, it can be found in readily available textbooks on physical oceanography.

The book begins with an account of the origins of beaches and the sources of beach sediment (Chapter 1), followed by the shaping of beaches and the changes that take place on them (Chapter 2), and then the causes of beach erosion (Chapter 3). The ways in which beaches have been influenced by artificial structures, such as sea walls and groynes, are dealt with in Chapter 4, and the principles and problems of beach nourishment (based on case studies that are well documented, or have been studied by the author) in Chapter 5.

Other management issues on beaches are introduced in Chapter 6. These include matters of ownership and jurisdiction, the provision of access, and the problems that result from various beach activities – problems such as litter and pollution on the shore and in nearshore waters. There are hazards on beaches that must be understood and dealt with, and there are ecological features associated with beaches that need to be managed in such a way as to maintain their value. Problems also arise from the diverse, sometimes conflicting, demands of the people who come to use the beach.

Just as there are variations around the world's coastline in the nature of beaches and the environments in which they have formed, so are there contrasts in beach use and behaviour between different countries and cultures. Nevertheless, many of the problems that require beach management are universal, and it is hoped that this book will provide a useful background to dealing with them.

I am grateful to Catherine Vinot, Jenifer Bird and Chandra Jayasuriya for help with the maps, Wendy Nicol for help with the photographs, and Juliet Bird for assistance in processing the text and preparing the index. The photographs are my own, except where otherwise indicated in the captions. My thanks are also due to people with whom I had discussions on beach management problems, notably Professor N. Psuty and his colleagues in the International Geographical Union's Commission on Coastal Systems.

Eric C.F. Bird
Oxford, May 1996

Chapter One

The origin of beach sediments

BEACHES

Beaches are unconsolidated deposits of sand and gravel on the shore. They fringe about 200 000 km (40%) of the world's coastline, the remainder being partly rocky, partly marshy or muddy, and partly artificial. There are many different kinds of beach. Some are long and almost straight or gently curved; others are shorter, and include sharply curved 'pocket beaches' in bays or coves between rocky headlands. Some are exposed to the open ocean or stormy seas, while others are sheltered in bays or behind islands or reefs; some are bordered by deep water close inshore, while others face shallow or shoaly water; some are fairly stable in plan and profile over periods of years or decades, others show rapid changes, especially in stormy weather. Some beaches are obviously gaining or losing sediment, some consist of sediment in transit (migrating along the coast), and others remain in position and may be relict, without any ongoing sediment supply. Most beaches show changes in plan (i.e. shape seen on a map or in a vertical air photograph) and in profile, either rapidly over periods of a few hours or days, or slowly over several decades or centuries.

The most notable variations are in the nature of beach sediments, and classification of beaches is generally based on their composition. Beach sediments consist of particles of various sizes (Table 1), the proportions of which can be determined by grain size (granulometric) analysis, as shown in Fig. 1. Some are coarse, dominated by cobbles and pebbles; others finer, with various grades of sandy sediment. Some are uniform (i.e. well-sorted), granulometric analysis showing a high proportion of a particular size grade; others are more varied in texture.

Most of the world's beaches are sandy, but coarser particles (gravels, which comprise granules, pebbles or cobbles, generally of stone but sometimes shelly) are often present, and may be scattered across a sandy beach, or arranged in patterns such as cusps or ridges running parallel to the shore. Sand and gravel particles may be angular or subangular in shape, but usually become rounded as they are worn down by the action of waves

Table 1 Beach grain size categories

Wentworth scale category	Particle diameter (mm)	ø scale
Boulders	>256	Below −8ø
Cobbles	64–256	−6ø to −8ø
Pebbles	4–64	−2ø to −6ø
Granules	2–4	−1ø to −2ø
Very coarse sand	1–2	0ø to −1ø
Coarse sand	½–1	1ø to 0ø
Medium sand	¼–½	2ø to 1ø
Fine sand	⅛–¼	3ø to 2ø
Very fine sand	1/16–⅛	4ø to 3ø

The Wentworth scale of particle diameters. The ø scale is based on the negative logarithm (to base 2) of the particle diameter in millimetres [$ø = \log_2 d$], so that coarser particles have negative values.

agitating the beach. Beaches composed entirely of well-rounded pebbles and cobbles are known as shingle beaches, especially in Britain, and there are also boulder beaches, with heaps of stones ranging in diameter up to more than a metre. Conventionally, deposits of sediment finer than sand (i.e. silt and clay) are regarded as muddy shores rather than beaches.

The term shore (or foreshore) is defined as the zone between highest and lowest tide levels, backshore as the zone above highest tide level, and nearshore as the zone between the water's edge (i.e. the shoreline, migrating to and fro with the tide) and the line where waves begin to break. The nearshore zone comprises the surf zone (with broken waves) and the swash zone (covered as each wave runs up the foreshore). The term coastline indicates the land margin (within the backshore zone), although in American literature the term shoreline is often used in this sense.

Beaches are readily mobilised by waves and currents, so that their outlines, both in plan and profile, change frequently. In general the coarser the beach material, the steeper the beach slope (Table 2). Winds also contribute to the shaping of sandy beaches when they are strong enough to move sand grains along or across the beach, onshore winds delivering sand to backshore dunes. Many beach profiles have an upper section, above normal high tide level, which is only occasionally submerged by large breaking waves or exceptionally high tides, a middle section, often more steeply sloping, between normal high and low tide levels, and a lower gently sloping section which is generally submerged even at low tides.

EVOLUTION OF BEACHES

During the past few million years the sea has risen and fallen several times around the world's coastline, and former coastlines (some with beaches)

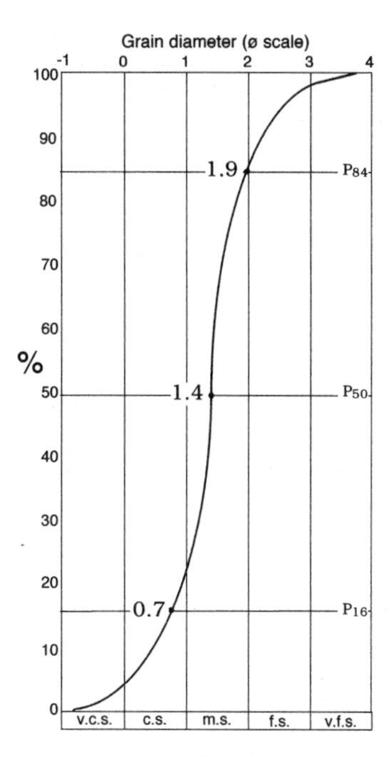

Figure 1 Granulometric analysis of a beach sample. The proportions of each grain size are measured by passing a dried sample of beach material of known weight through a set of sieves of diminishing mesh diameter, and so dividing it into size grades which are weighed separately. The results are usually presented as a graph of grain size distribution. Beach sediments are generally well sorted, which means that when they are analysed granulometrically the bulk of a sample falls within a particular size grade (fine sand in this example). The grain size distribution of a beach sediment is usually asymmetrical and negatively skewed, the mean grain size being coarser than the median. This is probably because the winnowing effect of wave (and on sandy beaches wind) action has reduced the proportion of fine particles.

In the simple example provided the median diameter (P_{50}, the 50th percentile) is 1.4, and the 16th and 84th percentiles are 0.7 and 1.9, respectively. Values for the mean, sorting (standard deviation) and skewness can be estimated as follows:

$$\text{Mean} = \tfrac{1}{2}(P_{16} + P_{84}) = \tfrac{1}{2}(0.7 + 1.9) = 1.3$$

$$\text{Sorting} = \tfrac{1}{2}(P_{84} - P_{16}) = \tfrac{1}{2}(1.9 - 0.7) = 0.3$$

$$\text{Skewness} = \frac{\text{Mean} - \text{Median}}{\text{Sorting}} = \frac{(1.3 - 1.4)}{0.3} = -0.33$$

can be found at various levels above and below present mean sea level. Major oscillations accompanied the waxing and waning of glaciers and ice sheets in polar and mountain regions during the Pleistocene Ice Age. The sea fell to lower levels during each phase of colder (glacial) climate, when the world's ice cover became more extensive and a larger proportion of the world's water was retained in the ice cover, and rose during the intervening

Table 2 Beach grain size categories and mean beach face gradients

Wentworth scale category	Mean slope of beach face (°)
Cobbles	24
Pebbles	17
Granules	11
Very coarse sand	9
Medium sand	7
Fine sand	5
Very fine sand	1

(interglacial) phases, when the melting of the ice fed water back to the oceans. At times the sea stood a few metres above its present level, and the beaches that then formed can be seen on some coasts as emerged or 'raised' beaches, usually beyond the present limits of wave action, even at high tide. During the last glacial phase, beginning about 80 000 years ago, global sea level fell about 140 m (Pirazzoli 1986), and the continental shelves were exposed as wide coastal lowlands.

About 18 000 years ago, in Late Pleistocene times, the climate began to become milder, and extensive melting of glaciers and ice sheets resulted in a world-wide sea level rise, the Late Quaternary marine transgression, also known as the Flandrian transgression because it was first documented in Flanders, in north-eastern France. The sea rose, probably with minor pauses and oscillations, submerging the continental shelves and advancing towards what is now the coastline. The marine transgression continued into Holocene times (defined as the past 10 000 years) and came to an end about 6000 years ago, when the general outlines of the world's coasts were established. On some coasts there have been further changes of sea level, largely because of uplift or subsidence of the land margin: the northern coasts of Canada and Scandinavia have been rising, so that the sea has there fallen relative to the land, whereas the Gulf and Atlantic coasts of the United States have been generally subsiding, so that the marine transgression has there continued more slowly through the Holocene to the present time. During the past 6000 years some coasts have receded as the result of marine erosion and cliffing, while others have advanced by the deposition of such features as deltas and coastal plains, and the formation and widening (progradation) of beaches.

Existing beaches are thus geologically of recent origin, formed as the Late Quaternary marine transgression slackened or gave place to a Holocene sea level stillstand. Some remain narrow, fringing cliffs and steep coastal slopes or bordering alluvial plains and wetlands, while others have widened with the addition of successively formed backshore beach ridges, or dunes built of sand winnowed from the shore. Beaches are rare on steep and rocky coasts, particularly where cliffs plunge into deep nearshore

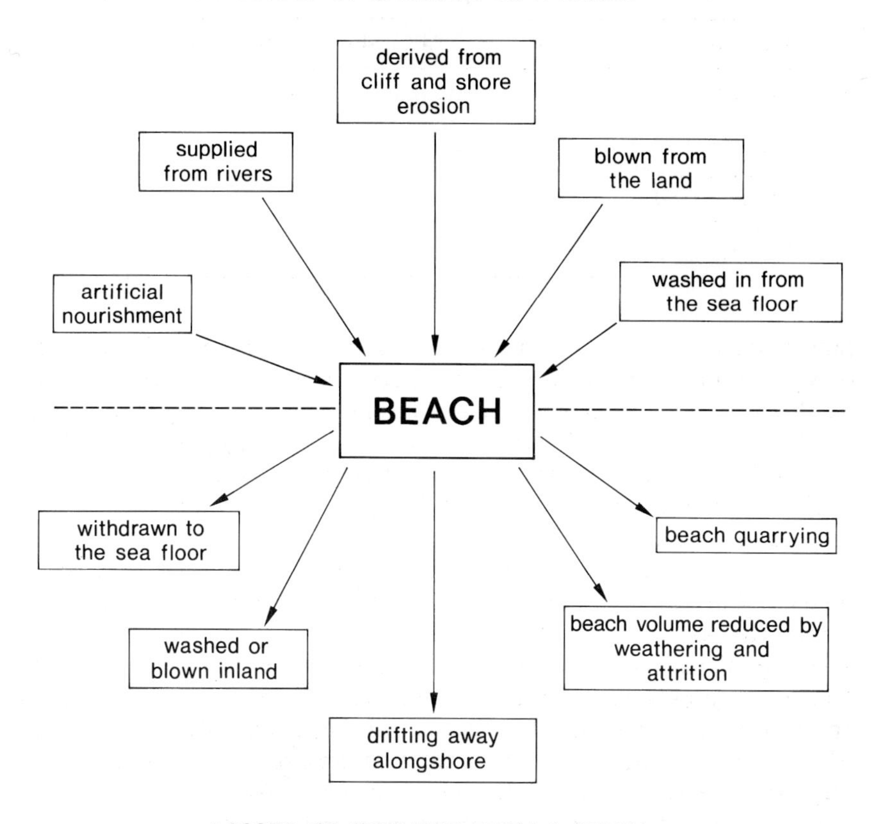

Figure 2 Gains and losses of beach sediment

water, and on shores dominated by silt or clay deposition, where they give place to salt marshes or mangrove swamps, usually behind mudflats exposed when the tide falls.

PROVENANCE OF BEACH SEDIMENTS

Beaches have received their sediments from various sources (Fig. 2). Some have been supplied with sand and gravel washed down to the coast by rivers. Others consist of material derived from the erosion of nearby cliff and foreshore outcrops, washed in from the sea floor by waves and currents, or delivered by winds blowing from the hinterland. In recent decades many beaches have been augmented by the arrival of sediment produced as the result of human activities, such as agriculture or mining on

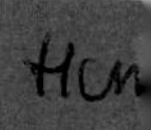

the coast or in the hinterland. Some beaches have been artificially nourished or replenished, especially at seaside resorts. While many beaches are still receiving sediment from one or more of these sources, some have become relict, and consist of deposits that accumulated in the past, but are now no longer arriving. The following sections describe these categories in more detail.

Beaches supplied with fluvial sediment

Fluvial nourishment of beaches occurs after sediment has been washed down to the mouth of a river (Fig. 3). Sand and gravel may then accumulate on the shores of a symmetrically growing delta, or be distributed alongshore in either direction by waves and currents to form beaches that can extend several kilometres along the coast. The coarser sand and gravel often remain on beaches near the river mouth, while the finer sediment is carried further along the shore. Rivers that flow into inlets, estuaries or lagoons may relinquish their loads of sand or gravel within these, rather than directly supplying a beach.

The nature of sediment supplied to beaches by rivers depends on several factors. It is determined by the types of rock that outcrop within the catchment, and in particular the material formed on the surface of these outcrops as they are decomposed or disintegrated by weathering processes. The volume and calibre of fluvial sediment loads are influenced by the steepness of the hinterland, the vigour of runoff produced by rainfall or the melting of snow or ice, the effects of earthquakes and volcanic eruptions, and the extent and luxuriance of the vegetation cover as modified by agriculture, forestry, mining and urban development. Examples of each of these are given below.

Sediment supplied by rivers to the coast has a mineral composition which reflects the characteristics of rock formations outcropping within each catchment. Weathered granite and sandstone produce sandy material, often dominated by quartz, the most durable of common rock minerals. Felspars and micas are less durable, but may persist in some beach sediments, especially those receiving abundant material from steep hinterlands. There are varying proportions of heavy (ferromagnesian) minerals in material derived from such rocks as granite, and these may become concentrated in beach sediments as layers of mineral sand, such as the rutile and ilmenite found on parts of the Australian coast, magnetite in southern Japan, and gold in Alaska. Sand brought down from volcanic hinterlands also reflects the composition of the source rocks: weathered basalt, for example, yields grey or black sand dominated by the dark mineral olivine.

In the British Isles there are few fluvially nourished sandy beaches, but sand from the Loire and the Gironde rivers has been delivered to beaches in western France, and there are fluvially fed sandy beaches around the

Figure 3 Supply of fluvial sediment to beaches, illustrated by the River Don, north-east Queensland, Australia. The sandy material in the channel is washed down to the river mouth during periods of strong fluvial discharge, and is then reworked and distributed along the coast by wave action

Mediterranean and on the north-west coast of the United States, notably those supplied by the Columbia River. Sandy beaches derived from sandstone catchments are extensive in Southern California, and on the south-eastern coast of Australia some beaches consist largely of fluvially supplied quartz sand originating from weathered granite. Examples of beaches composed primarily of fluvially supplied sand are also seen on coasts adjacent to the mouths of the Gascoyne, Ashburton and De Grey Rivers in north-western Australia, each of which has a large catchment of sandy desert from which substantial quantities of sand are occasionally carried downstream when rapid runoff is generated by the heavy rainfall that accompanies tropical cyclones.

Shingle beaches are found where rivers drain catchments with fissile rock outcrops or extensive gravel deposits and flow strongly enough to deliver stony material to the shore. In the North Island of New Zealand, Quaternary volcanic deposits include much coarse gravelly material, some of which is carried downstream by such rivers as the Mohaka, which has formed pebbly beaches along the north coast of Hawke Bay (Bird 1996).

Where the hinterland is mountainous, as on the west coast of South Island, New Zealand, the north coast of New Guinea bordering the Torricelli Mountains, or the Caucasian Black Sea coast, strongly flowing rivers carry gravel as well as sand down to the coast, to be incorporated in beaches. Gravels are also common in areas where past or present glaciation has produced large quantities of stony morainic debris which can be washed down to the coast, as in southern Alaska and south-eastern Iceland. Shingle beaches on the north coast of Scotland include gravelly material derived from Pleistocene glacifluvial deposits and brought down by the Rivers Spey and Findhorn.

Much of the sediment delivered to the mouth of a river arrives during episodes of flooding after heavy rainfall or the sudden melting of snow or ice in the river catchment. Floodwaters may deposit sediment on and around a delta, or carry it out to form an offshore shoal, which is reworked by waves and currents after the floods abate, some of the sand or gravel being washed onshore to be added to the beach. On steep coasts, such as those in southern Brazil, beaches are periodically renourished with sand brought down by river floods from a steep hinterland where soil erosion has generated an abundance of fluvial sediment (Cruz et al. 1985). In southern Alaska, streams swollen by spring meltwater have carried large quantities of sand and gravel from morainic deposits and glacial outwash down to the coast to form beaches on the shores of Icy Bay (Molnia 1977).

Earthquakes in the hinterland can increase fluvial sediment yields. The severe tremors which shook the Torricelli Mountains in northern New Guinea in 1907 and 1935 caused massive landslides which delivered large quantities of sand and gravel to rivers draining northward to the coast, where beaches were widened, and beach material subsequently drifted alongshore from river mouths, eastward to Cape Wom (Bird 1981). Volcanic eruptions within a river basin can also generate downstream movement of large quantities of sand and gravel. In the North Island of New Zealand, river loads have occasionally been increased as the result of episodes of volcanic activity. The Tarawera volcanic explosion in 1886 produced sands and gravels that moved down rivers such as the Rangitaiki, and beaches on the shores of the Bay of Plenty prograded when this material reached the coast (McLean 1978). In a similar way eruptions of Merapi volcano, north of Jogjakarta in Java, have resulted in downstream flow of sand and gravel, and the arrival of this sediment at river mouths has been followed by the progradation of beaches of grey volcanic sand at and west of Parangtritis (Bird and Ongkosongo 1980). In south-east Iceland glacifluvial streams have had their loads further increased by the eruptions of hinterland volcanoes, which cause catastrophic ice melting and river flooding, leading to the formation of wide outwash plains, known as sandur, the seaward fringes of which are subsequently reworked by wave action to form beaches (Bodéré 1979). However, although the Mount St. Helens eruptions in 1980 delivered vast quantities of sediment to the Columbia and other rivers flowing down to the Oregon coast, much of it was silt and clay, and there has been little ensuing progradation of the beaches.

Runoff from unvegetated slopes is much more rapid than from slopes that carry a plant cover which intercepts, retards and recycles rainfall. As a result, rivers draining catchments where the vegetation cover has been reduced or removed deliver larger and coarser sediment loads to the coast. Around the Mediterranean Sea, for example, deforestation, overgrazing and excessive cultivation of hinterlands has led to increased runoff, soil erosion and river flooding, so that larger quantities of fluvial sediment have reached the mouths of rivers, and beaches have prograded. Examples of this have been reported from Greece and Turkey, where deposition has been rapid at and around river mouths since classical times (Bird 1985a). The effect depends on the continued presence of weathered material on the hinterland slopes, for where this is removed entirely the fluvial sediment yield diminishes. The Argentina River, on the Ligurian coast of Italy, had a phase of delta growth and beach progradation when sediment yield from the steep, eroding catchment was high, but in recent decades the supply has declined as bedrock became widely exposed upstream (Bird and Fabbri 1993).

Some beaches have been enlarged by the accumulation of sand or brought down by rivers from mining areas in the hinterland. On the island of Bougainville, north of New Guinea, waste from copper mines increased the load of the Kawerong River, and has been carried downstream to be deposited on beaches at and around the river mouth at Jaba, on the shores of Empress Augusta Bay (Brown 1974). In New Caledonia extensive hill-top quarrying of nickel and chromium ores resulted in massive spillage of rocky waste into river valleys, and the delivery of increased and coarsened fluvial loads to beaches at and near river mouths, as at Karembé on the western coast and Thio and Houailou on the eastern coast (Bird et al. 1984). In Chile the arrival of sandy tailings washed downstream from copper mines resulted in progradation of the beaches in Chañaral Bay (Paskoff and Petiot 1990), and in south-west England tin and copper mining and the quarrying of china clay from Hensbarrow Downs, a granite upland north of St. Austell, produced large quantities of sand and gravel waste with much quartz and felspar, some of which was carried down rivers to be deposited on beaches at Par and Pentewan (Everard 1962).

The contribution of river sediment to beaches has thus been substantial in many parts of the world, but few beaches are entirely of fluvial origin and many have been nourished primarily from other sources, which will now be considered.

Beaches supplied from eroding cliffs and foreshores

Beach sediments derived from the erosion of cliffs and foreshores have characteristics related partly to the lithology of the outcropping formations and partly to the energy of the waves and currents which erode material from these outcrops and carry it along the shore. Sand eroded from cliffs cut in soft sandstone has nourished beaches, for example at Point Reyes on the coast of Southern California. Cliffs cut in glacial deposits have produced sand and gravel delivered to beaches around Puget Sound (Shepard and Wanless 1971) and on other similar coasts, as in New England, the Danish archipelago and the southern shores of the Baltic, notably in Poland. Beaches occur where glacial moraines intersect the coastline and cliffs cut into these deposits have provided a source of sand and gravel which has been spread along the coast: examples have been noted in southern Norway, Finland and Estonia (Orviku et al. 1995).

In general, gravel beaches are found where the coastal rock formations have yielded material of suitable size, such as fragments broken from thin resistant layers in sedimentary rocks, or intricately fissured formations, or pebbly conglomerates (Fig. 4). They are not found where the coastal rock outcrops are homogeneous, as on massive granites, or where they are soft and sandy.

Extensive shingle beaches have been derived from the nodules of hard flint eroded from chalk cliffs on the south coast of England and the north coast of France. As the cliffs are cut back the flint nodules released by weathering and erosion form gravelly beaches, which are agitated by wave action and soon become well-rounded shingle. This drifts eastward along the Channel coast to sectors of accumulation such as the cuspate foreland at Dungeness and the Pointe du Hourdel at the mouth of the Somme. On tropical coasts there are gravel beaches derived from the lateritic ironstone crusts that outcrop in coastal cliffs and rocky shores, as in the Darwin

Figure 4 Beach material derived from cliff erosion, illustrated near Budleigh Salterton, Devon, England, where wadi gravels from Triassic deposits in the receding cliff are incorporated in the cobble beach

district in Australia and Port Dickson in Malaysia. Sand and gravel beaches on the shores of Anak Krakatau, Indonesia, have been derived from the erosion of cliffs cut in unconsolidated volcanic ash and agglomerate (Bird and Rosengren 1984), and there are similar beaches around the modern volcanic island of Surtsey, which began to form off the south coast of Iceland in 1963 (Norrman 1980).

Some beaches have incorporated material derived from Pleistocene 'raised beaches' that formed above present sea level along cliffed coasts during phases of higher sea level, as in Falmouth Bay, Cornwall. Others have been supplied by landslides which form protruding lobes, usually with much silt and clay, as on the shores of Lyme Bay and on the North Norfolk coast in England. As these are consumed by wave action the finer sediment is dispersed, and the sand and gravel fractions are released to be added to nearby beaches.

If the sand or gravel eroded from cliffs accumulates as basal talus, erosion slackens and the cliffs become degraded bluffs. The supply of cliff-derived sediment to beaches is thus reduced. A continuing sediment supply from cliff erosion can occur only where waves move the fallen material away along the coast, or where it is being carried offshore. Usually the beach material eroded from a cliffed sector accumulates nearby, in an embayment or on a spit or foreland.

Beaches supplied with sediment from the sea floor

Sediment washed in to beaches from the sea floor includes sand or gravel eroded from submerged geological outcrops or collected from unconsolidated bottom deposits. Some of these originated from the hinterland, and were spread across the continental shelf by river outflow, glaciers or wind action during the several phases of lower sea level in Pleistocene times. Just as there are river catchments on land from which coastal sediment has been derived and delivered to the coast at river mouths, so are there 'sea floor catchments', areas of the continental shelf from which beach material has been, and may still be, carried shoreward, and delivered to parts of the coast.

Shoreward drifting of sediment has occurred where the nearshore waters are shallow, or are becoming shallower because of land uplift or a falling sea level. This can be seen in the Gulf of Bothnia, where isostatic uplift is causing emergence, so that sand and gravel from emerging eskers and morainic deposits are being swept shoreward on to beaches, as at Storsand in Sweden, Brusand in Norway and Kalajoki in Finland. Similar emergence has led to the formation of a wide sandy beach on the south-western shore of the island of Laesø in the Kattegat (Møller 1985). Shoreward movement of sea floor sand took place during the phase of lowering of the Caspian Sea between 1930 and 1977, when beaches were widened by accretion as the shores emerged, and around the Great Lakes similar beach progradation has occurred as the result of shoreward drifting during each of several phases of falling lake level (Olson 1958, Dubois 1977).

Widespread shoreward drifting took place on many coasts, particularly oceanic coasts, during the Late Quaternary marine transgression (p. 4) as the sea advanced across shoaly topography and waves washed sediment on to the shore, forming beaches and dunes. As the sea rose some of the beach material was washed or blown landward, and it is probable that minor oscillations of sea level occurred, facilitating shoreward sweeping of sand and gravel on to beaches. Chesil Beach is an example of a shingle barrier that formed and was driven shoreward as the sea rose, with a lagoon (The Fleet) on its landward side backed by a submerging, indented hinterland coastline that was never exposed to the open sea (Steers 1953).

Shoreward drifting of sand and gravel has produced beaches in suitable niches, such as shallow bays, coves and inlets on steep and rocky coasts. On oceanic coasts swell waves have shaped long gently-curving beaches, often backed by wide sandy plains with multiple beach ridges and dunes that formed after the Late Quaternary marine transgression came to an end. On the Gippsland coast in south-eastern Australia, where most of the rivers deposit their loads in estuaries or lagoons (such as the Gippsland Lakes) behind wide dune-capped Holocene coastal barriers, and where eroding cliffs are very limited, the Ninety Mile Beach has been formed almost entirely of sand swept in from the sea floor during the Late Quaternary marine transgression and the ensuing stillstand. Successively formed beach or dune ridges on the coastal barriers backing the modern beach are evidence of Holocene progradation, which in places is still continuing. At the south-western end of the Ninety Mile Beach sand is still moving shoreward on to the beach, and there is similar shoreward movement of sand from shoals in Streaky Bay in South Australia (Bird 1978).

Other examples of coasts where sand from nearshore shoals is still being carried onshore to be added to beaches include south-western Denmark, where the sandy islands of Fanø, Mandø and Romø have beaches prograding in this way, parts of the southern coast of Florida, and near Montevideo in Uruguay. Sand is also being washed in from a shallow sea floor to prograde beach sectors on the shores of Carmarthen Bay in South Wales, Holy Island in Northumberland, Studland in Dorset and Tentsmuir in Scotland (Bird and May 1976). However, on many coasts progradation of beaches by the arrival of sand from the sea floor has come to an end, the available supply of sand having diminished over recent centuries (p. 77).

The most obvious deposits originating from the sea floor are sand or gravel derived from marine organisms, notably shells, while on tropical coasts there are beaches of coralline gravel derived from the disintegration of fringing and nearshore coral reefs. Some shelly beaches are composed of rock-dwelling species while others have come from shallow sandy and muddy environments in sheltered bays and estuaries (Gell 1978). There are mussel shell beaches bordering the Sea of Azov, sandy beaches composed of oolites formed in the clear warm seas of the Bahama Banks in the Atlantic, and shell grit beaches on the shores of the Hebrides in western Scotland. Shelly beaches are also common on the shores of estuaries and lagoons.

As a rule, shelly beaches are mixed with varying proportions of inorganic sand or gravel, but beaches bordering low-lying (i.e. uncliffed) desert coasts are often strongly calcareous because the meagre runoff has yielded very little fluvial terrigenous sediment supply to the shore, so that beaches are dominated by material washed in from the sea floor. The Eighty Mile Beach, in north-western Australia, exemplifies this, and there are shelly beaches bordering the dry northern shores of Spencer Gulf near Whyalla in South Australia. In southern and western Australia there are calcareous sandy beaches consisting of shelly debris, and material from sand-sized organisms such as foraminifera and bryozoa carried in from the continental shelf by wave action (Bird 1978).

If other sources of natural beach nourishment diminish, it is likely that the proportion of biogenic sediment in a beach will increase on coasts where nearshore waters provide a rich environment for shelly organisms.

Beaches supplied with wind-blown sand

Onshore winds blow sand to the back of a beach, where dunes may be formed, and sand can also be swept alongshore by wind action. Some beaches have been supplied with sand blown from the hinterland, if there is a suitable source of unconsolidated sand with little or no retaining vegetation, and winds blow from the land to the sea. This happens on arid coasts, as in Angola, where barchans spilling on to the shores of Tiger Bay have added sand to local beaches (Guilcher 1985). Other examples have been reported from the desert coasts of Namibia between Sandwich Harbour and Conception Bay (Bremner 1985), Mauritania (Vermeer 1985), the Bahia de Paracas in Peru (Craig and Psuty 1968), and in south-eastern Arabia (Sanlaville 1985).

Where dunes have been built by onshore winds, sand may occasionally be swept back to the beach, and into the sea, by winds that blow from the land. An example of this is seen on the north-facing shores of the Slowinski National Park, on the Polish Baltic coast, where a wide beach has been partly nourished with sand blown by southerly or south-westerly winds from poorly vegetated backshore dunes (Borowca and Rotnicki 1994).

Where the prevailing winds blow more or less alongshore, dunes may drift across promontories and headlands to nourish beaches on the lee coast. There are examples of this on the south-facing Cape Coast of South Africa, notably at Port Elizabeth, where sand driven by the prevailing westerly winds across Cape Recife has been spilling on to the eastern beaches, and has then drifted alongshore to prograde King's Beach, beside the harbour (Heydorn and Tinley 1980). There are similar situations on the Victorian coast in south-eastern Australia, notably at Cape Woolamai (Fig. 5) and on the Yanakie Isthmus, Wilsons Promontory (Bird 1993a), and in Uruguay, where dunes are spilling on to the shore and nourishing the beach near Castillos (Jackson 1985).

Beaches made or modified by human activities

Many beaches contain small proportions of sand or gravel formed from fragments of glass, concrete, brick and earthenware, which are the product of human activities. These are more prominent on coasts with a long history of human settlement, as around the Mediterranean, and close to large urban or industrial centres. At Workington, on the coast of Cumbria, north-west England, the beach is dominated by basic slag waste from a former steel works. While the steel works was active, dumping of this material prograded the coastline, comparison of 1884 maps with 1981 air photographs showing an advance of up to 200 m. The slag tip has since been cliffed by marine erosion, and as the cliffs retreat, sand and gravel derived from the waste drifts northward alongshore, augmenting the

Figure 5 Supply of wind-blown sand (arrowed) to a beach on the shores of Cleeland Bight, Phillip Island, Australia. See Fig. 38

natural beach as far as the Pell Mell breakwater at the mouth of the Derwent River (Empsall 1989).

Reference has been made to beaches supplied with fluvial sediment augmented by mining activities in the hinterland (p. 9), but some beaches have been directly augmented by material from quarry waste spilling on to the shore. At Hoed, in Denmark, wave action has distributed gravelly flint waste dumped from a coastal quarry, and built it into a series of low parallel beach ridges fronted by a shingle beach (Fig. 6) (Bird and Christiansen 1982). Gravelly waste from coastal quarries, tipped over cliffs into the sea on the east coast of the Lizard Peninsula, in south-west England, has accumulated in adjacent coves at Porthallow and Porthoustock as widened gravelly beaches (Fig. 7) (Bird 1987), and the beach at Rapid Bay in South Australia has also been formed from spillage of quarry waste (Bourman 1990). On the west coast of Corsica the arrival of waste debris from an asbestos mine produced a beach in a cove where none existed previously (Paskoff 1994). Other examples include the Nganga Negara tin mine site on the west coast of peninsular Malaysia, where the dredging of tin from a depositional apron and coastal plain fronting the granitic Segari Hills generated large quantities of quartzose sand, gravel and boulders, which were heaped as a high tailings bank along the coastal fringe. Waves have reworked this material to form a gently shelving sandy beach, finer and better sorted than the tailings sediment. Waste material from the large

Figure 6 Dumping of gravelly quarry waste on the shore at Hoed, north-east Jutland, Denmark, has resulted in the formation of a wide beach

granite and diorite quarry at Ronez, on the north coast of Jersey, has spilled down to the shore and formed a beach downdrift to the east.

Waste dumped on the shore from coal mines has developed into a beach of black pebbles and sand at Lynemouth in Northumberland, and similar material from coastal collieries has been dumped on several sectors of the Durham coast, notably between Seaham and Easington and at Horden (Carter 1988). This has widened the adjacent beaches, and sediment has drifted alongshore from these to augment beaches southward to Hartlepool. On Brownsea Island, in Poole Harbour, southern England, wave action has distributed debris from a former pipeworks to build a beach of broken subangular earthenware.

Driftwood, including sawn timber, is extensive on beaches bordering high forested hinterlands, as in British Columbia. On the beaches of the

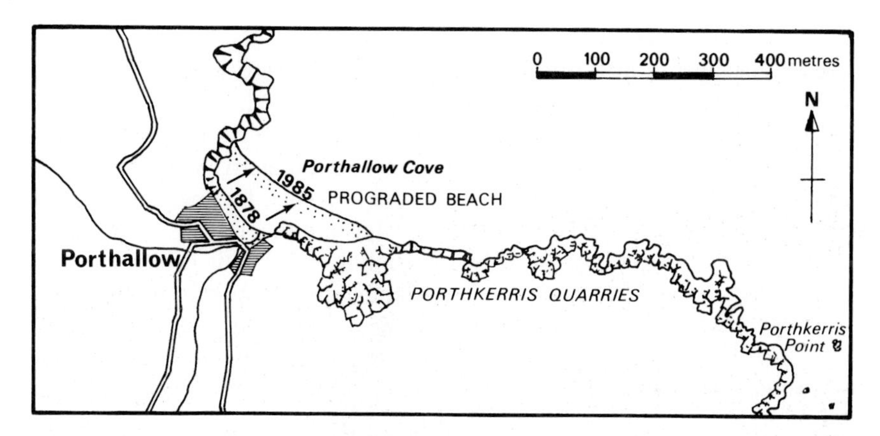

Figure 7 Progradation of the beach in the cove at Porthallow, Cornwall, was due to the arrival of gravelly material derived from waste tipped into the sea from the nearby Porthkerris Quarries

Westland coast, New Zealand, the driftwood consists mainly of trees washed down to river mouths, but coasts adjacent to lumbering areas in western Canada and northern Russia are heaped with sawn timber washed up on the shore. On eroding sandy coasts in south-eastern Australia many beaches are littered by branches and twigs from undercut trees. On some beaches driftwood and lumber piled on the shore can act protectively by impeding wave scour, but it is also possible for fallen trees and branches to be jostled by wave action, and to act as levers or battering rams, thereby helping to erode sand or gravel from the beach.

Many beaches contain material derived from garbage of various kinds, including fragments of broken glass, metal, brick and plastic, derived from bottles, cans, containers and other litter dumped on the shore, carried down by rivers, or washed in from the sea. The origin of beach litter is discussed in Chapter 6 (p. 225).

Nourished beaches, formed partly or wholly by the dumping of sand, gravel or similar material, are now extensive, particularly in the United States, Europe and Australia. These will be discussed in Chapter 5.

Beaches of mixed origin

As has been noted, many beaches contain sand or gravel from more than one source. In Southern California, for example, beaches have been largely fed with fluvial sand from such rivers as the Santa Clara and Ventura, but they also include material eroded from coastal cliff outcrops, as well as sediment, including shelly debris, washed in from the sea floor (Emery 1960). On the Cape Coast of South Africa the beaches include sand of aeolian origin where dunes are spilling along the shore, as well as fluvial, marine and cliff-derived sediment (Heydorn and Tinley 1980). Fluvially supplied beaches downdrift from a river mouth may gradually become mixed with sediment from other sources, such as eroding cliffs or the sea floor. The fluvial contribution may remain identifiable from mineralogical evidence. Thus the Ninety Mile Beach in south-eastern Australia, which consists largely of well-rounded quartz sand and shell fragments washed in from the sea floor, is, in the vicinity of the mouth of the Snowy River, mixed with fluvial sand, supplied by occasional river floods, which contains distinctive minerals, including augite (McLennan 1976). In north Queensland, beach sediments are varying mixtures of sand from rivers, eroding cliffs or nearshore reefs: samples from Garners Beach, near Bingil Bay, were found to contain quartz and felspar sand supplied by nearby rivers, lithic fragments from local rock outcrops, particles of disintegrated beach rock (p. 39), ferruginous sandrock from an eroded Pleistocene sandy barrier, and coralline and shelly material from an adjacent reef (Bird 1971). On the western shores of the Arabian Gulf the beaches consist partly of sand blown from the desert, but also include shelly and inorganic sediment washed up from the sea floor (Sanlaville 1985). Pumice derived from submarine volcanic eruptions floats on sea water, and is a common, and often far-travelled, constituent of many oceanic beaches, particularly around the Pacific.

Relict beaches

Some beaches have become relict because the sediment sources that originally supplied them are no longer available. This may be due to the natural or artificial diversion of a river, the construction of a dam, or the implementation of successful anti-erosion works in a catchment, so that the former fluvial supply of sediment to the beach has ceased. Alternatively, it could be the result of the halting of cliff erosion by sea wall construction or emergence due to land uplift or sea level lowering, resulting in withdrawal of wave attack from the cliff base. On many oceanic beaches there is evidence that sediment is no longer being washed in from the sea floor because the supply from unconsolidated shoaly deposits has run out (p. 77).

Relict beaches, continually reworked by wave action, can become very well sorted, or develop lateral grading in grain size in relation to incident wave regimes. Such adjustments are impeded where there is a continuing supply of fresh sediment. There is also a tendency for the grain size of beach sediment to diminish as the result of gradual attrition by wave action. Slapton Beach, in Devonshire, England, is an example of a well-sorted relict beach of coarse sand and fine shingle, the calibre of which has been reduced by prolonged attrition resulting from wave action. Some have considered Chesil Beach, on the eastern shores of Lyme Bay, to be essentially a relict beach because it is so well sorted, with a remarkably good lateral gradation from small pebbles in the west to large cobbles in the south-east, but this beach may still be receiving some shingle from along-shore sources (p. 67).

BEACH ACCRETION AND EROSION

Beaches that receive more sediment from the various listed sources than they lose onshore, alongshore or offshore are built upward and outward, so that the high and low tide lines advance seaward, and the coast progrades. Beach erosion occurs where the losses of sediment exceed the gains. It is now necessary to consider the processes that result in the evolution and dynamics of beaches.

Chapter Two

Beach processes and morphology

This chapter examines the changes that take place on beaches in terms of the various nearshore processes at work on them. It begins with a brief description of nearshore processes as background for the study of beach management. If more detail is required on these processes it is available in textbooks such as those by King (1972), Komar (1976), Bird (1984), Pethick (1984), Carter (1988), Hardisty (1990) and Viles and Spencer (1995), as well as in the Schwartz *Encyclopedia of Beaches and Coastal Environments* (1982). Short (1996) has edited a recent work on beach morphodynamics.

COASTAL PROCESSES

Introduction

The processes at work in coastal waters include waves, tides and currents, which together provide the energy input which shapes and modifies beaches by eroding, transporting and depositing beach sediments. Although waves, tides and currents interact, one process sometimes augmenting or diminishing the effects of another, it is convenient to treat them separately at first.

Waves

Waves are undulations on a water surface produced by wind action. The turbulent flow of the wind blowing over water produces stress and pressure variations on the surface, initiating waves which grow as the result of a pressure contrast that develops between their driven (upwind) and advancing (downwind) slopes. Waves consist of orbital movements of water that diminish rapidly from the surface downwards, until the motion is very slight where the water depth (d) equals half the wave length (L). The depth at which waves become imperceptible is termed the wave base, and in theory wave erosion could ultimately reduce the world's land areas to a planed-off surface at this level. The orbital motion in waves is not quite complete, so that water particles move forward as each wave passes, producing a slight drift of water in the direction of wave advance. Wave height (H) is the vertical distance between successive crests and troughs, wave steepness the ratio between height and the length (H/L), and wave velocity (C) the rate of movement of a wave crest.

Wave dimensions are determined partly by wind velocity, partly by fetch (the extent of open water across which the wind is blowing), and partly by the duration of the wind. In coastal waters where fetch is limited, the height of the waves is proportional to wind velocity, and the wave period (T, the time interval between the passage of successive wave crests) is proportional to the square root of wind velocity. In mid-ocean the largest waves, generated by prolonged strong winds over fetches of at least 500 km, can be more than 20 m high, and can travel at more than 80 km/h. In deep water, wave velocity (C, the rate of movement of a wave crest in metres per second) is the ratio (L/T) of wave length (measured in metres) to wave period (measured in seconds).

It may be useful to give some elementary equations illustrating the relationships between wave parameters*. Wave length (L_o) in deep water (where $d > L/2$) can be used to calculate wave velocity (C_o) from the following formula, in which g is the gravitational acceleration (about 980.62 cm/s^2 at latitude 45°):

$$C_o^2 = \frac{gL_o}{2\pi}$$

from which, since $L_o = C_oT$,

$$C_o = \frac{gT}{2\pi} \text{ or } 1.56T \text{ in m/s}$$

so that $L_o = 1.56T^2$, a convenient way of calculating wave length from measurements of wave period in deep water.

Wave trains

Some waves in the nearshore zone are generated locally by winds blowing over coastal waters, particularly onshore winds, but on coasts exposed to a long fetch (especially ocean coasts) there are also waves of distant derivation, transmitted from remote storm centres, which arrive as swell. During storms winds generate irregular patterns of waves, varying in height, length and direction, which radiate from the generating area. The longest waves move most rapidly and are most durable, so that the travelling wave train

---

* More detail on these and other equations dealing with wave action and its effects can be found, for example, in Komar (1976) and Carter (1988).

becomes sorted into swell of more regular height and length, which may travel far across the oceans and eventually arrive on some distant beach.

Ocean swell consists typically of long, low waves with periods of 10 to 16 s; the wave crests gain in height and steepness as they enter shallow water, and break to produce the surf that washes on to beaches, as on the shores of the Pacific, Atlantic and Indian Oceans. The most prolific swell-generating region is the Southern Ocean, where storms initiate the south-westerly swell that arrives on the beaches of Australia, New Zealand and South Africa, and is transmitted into the Indian Ocean to the coasts of southern Asia, across the Pacific to the coasts of Mexico and California, up to Alaska, and across the Atlantic to West Africa and Western Europe. Occasionally waves generated in the vicinity of the Falkland Islands reach the south-west coasts of Britain as ocean swell (wave periods up to 20 s) breaking on the Loe Bar in Cornwall or Chesil Beach in Dorset.

Storms in northern latitudes generate similar ocean swell, especially in winter, when a north-westerly swell from the north Pacific reaches the beaches of California and Central America, and in the Atlantic a similar swell extends to the coasts of Portugal and West Africa. There are seasonal alternations, a winter north-westerly swell alternating with a summer south-westerly swell (i.e. stronger in the Southern Ocean winter) reaching the coasts of Portugal and California, and producing the seasonal contrasts seen, for example, at Half Moon Bay, near San Francisco, California.

Waves generated by wind action in coastal waters are typically shorter (wave period less than 10 s) and less regular than ocean swell of distant derivation, and in stormy periods they are much steeper. They may be superimposed on ocean swell arriving in coastal waters, an onshore wind accentuating the swell and adding shorter waves to it, a cross wind producing shorter waves which move at an angle through the pattern of swell, and offshore winds flattening ocean swell to produce relatively calm conditions in the nearshore zone.

Nearshore waves

Sectors of coast protected by promontories, reefs or offshore islands receive swell (if at all) in a much modified and weakened form, so that locally wind-generated waves predominate on beaches. This is the case around landlocked seas, such as the Mediterranean and the Baltic, the Arabian Gulf and the Gulf of California, and in embayments with constricted entrances (such as Port Phillip Bay, Australia). Around the British Isles wave regimes are largely determined by winds in coastal waters.

Geographical variation in nearshore wave conditions has been explored by Davies (1980), whose diagram of world coastal wave environments has been reproduced in many textbooks. Variations in the direction of ocean swell were noted in the previous section, but the stormiest sectors of ocean coast are the west-facing coasts of Europe, Canada and Patagonia, in latitudes subject to frequent westerly gales. Waves are generated by occasional tropical cyclones (also known as hurricanes or typhoons), most frequently in the Caribbean, the north and west Indian Ocean, the China Sea and northern Australia, but the coastal waters of equatorial regions (such as north-eastern Brazil and Indonesia) are relatively calm except where they receive ocean swell of distant derivation, as in the Gulf of Guinea, southern Sumatra and the Pacific coast of Central America. In Australia, much of the coast is dominated by ocean swell, but in Queensland, where barrier reef structures offshore largely exclude such swell, and on the shores of the Gulf of Carpentaria waves are generated by the prevailing south-easterly trade winds.

Wave refraction

Ocean swell has parallel wave crests in deep water, but as the waves move into shallower water they begin to be modified by the sea floor: the free orbital motion of water is impeded, and the frictional effects of the sea floor retard the advancing waves. Sea floor topography thus influences the pattern of swell moving in to nearshore waters, the angle between the swell and the submarine contours diminishing, so that the wave crests become realigned until eventually they are parallel to the submarine contours.

This is known as wave refraction. Where the angle between the swell and the submarine contours is initially large the adjustment is often incomplete by the time the waves arrive at the beach, so that they break at an angle (usually less than 10°). Where the angular difference is initially small, the waves are refracted in such a way that they anticipate and fit the outline of the beach, and break synchronously along its length (Fig. 8). Sharp irregularities of the sea floor have stronger effects: a submerged bank retards the advancing waves, but a submarine trough allows them to run on. Islands or reefs awash at low tide produce complex patterns of wave refraction, and waves that have passed through narrow straits or entrances are modified by diffraction, spreading out in the water beyond. Waves are also retarded as they impinge on headlands, and run on into the intervening bays. In broad embayments they show gently curved patterns, the wave in the middle of the bay moving on in deeper water while towards the sides, in shallower water, it is held back. In narrow embayments the refracted wave crests become more sharply curved.

Patterns of wave refraction in coastal waters can be seen on vertical air photographs, and methods have been devised for determining them graphically (CERC 1987) or by use of a computer programme (e.g. Brampton 1977). Given knowledge of the direction of approach of waves in deep water, their length or period, and the detailed configuration (bathymetry) of the sea floor from nautical charts, it is possible to construct diagrams

Figure 8 The curved outline of the beach in Seven Mile Bay, Tasmania, has been shaped by gently refracted ocean swell, breaking to form surf which sweeps sand on to the shore. A swash-dominated beach (see p. 44)

showing the patterns of refraction that will develop as waves enter nearshore waters and approach the coast. The modifications can also be demonstrated by drawing orthogonal lines (rays) at equal intervals, perpendicular to the alignment of waves in deep water, and running shoreward. These rays converge where they pass over a submerged bank or reef, and diverge where deeper water is traversed; in general, they converge towards headlands and diverge in embayments. The spacing of such orthogonals shows how wave energy is distributed, convergence towards a section of beach indicating a concentration of wave energy, whereas divergence indicates a weakening. This can be expressed as a refraction coefficient (R), equivalent to the square root of the ratio between the distance between neighbouring orthogonals in deep water (S_o) and their spacing on arrival at the shoreline (S). Calculated from measurements on a wave refraction diagram, this coefficient gives an approximate expression of the relative wave energy arriving on each sector of the beach.

Wave energy

There are relationships between patterns of refracted waves approaching the shore and the morphology and sediment characteristics of beaches. On sectors where convergence of orthogonals indicates augmented wave energy (i.e. larger waves breaking on the beach) beaches become generally steeper and higher, with coarser grain size, better sorting, more severe erosion, and divergence of alongshore currents causing sediment dispersal. Divergence of orthogonals, indicating low wave energy sectors, shows the reverse of these conditions: lower beaches with gentler gradients, generally finer and less well-sorted beach sediment, reduced erosion or perhaps accretion, and convergent longshore currents bringing in beach sediment. The relationships may be complicated, however, by other factors, such as the nature of available sediment: there are many sandy beaches on high wave energy coasts where coarser sediment is unavailable, and gravelly beaches do occur on low wave energy coasts where there is a source of such material. Positions of such features as stream outlets or lagoon entrances are related to sectors of diminished wave energy, as indicated by wider spacing of orthogonals.

Beach outlines in plan are thus related to patterns of approaching waves, and wave refraction diagrams can also be used to predict the direction that longshore currents will develop when waves arrive at an angle to the shore.

Breaking waves

As waves move into shallow water ($d < L/2$) they are modified in several ways. Their velocity C_s diminishes according to the formula

$$C_s^2 = \frac{gL}{2\pi.\tanh 2\pi d/L}$$

The shallowing water also diminishes their length and period, and as they approach a beach their height increases. The waves steepen, the crests becoming narrower and sharper, and the intervening troughs wider and flatter. Orbital movements within each wave therefore become more and more elliptical, and shoreward velocity in the wave crest increases until it exceeds the wave velocity. When the orbital motion can no longer be completed, the oversteepened wave front collapses, forming a breaking wave (breaker). This sends forth a rush of water through the surf zone and thence, as swash, up on to the beach, followed by a withdrawal, known as the backwash, by which water returns to the sea.

The pattern of breaking waves varies with wave height and nearshore gradient, so that low waves break closer to the shore than high waves, and waves of a specific height break closer inshore where the gradient is steep than where it is gentle. There are complications where bars are formed by wave action in the nearshore zone: the number, spacing and amplitude of the bars varies with the height and period of breaking waves, but the bar topography, in turn, influences where and how incident waves will break.

The ways in which waves shape and modify a beach depend on their incidence and dimensions as they reach the shore, how they break, and the patterns of water flow that result. These features are influenced by the effects of local winds, by changes in nearshore water depth accompanying the rise and fall of tides or other short-term sea level changes, by currents other than those resulting from wave motion, by the gradient and topography of the sea floor, and by the general configuration of the coastline.

The energy of a breaking wave (E) depends largely on its height (H_b) and the density of the water (p), together with gravitational acceleration (g), according to the formula

$$E = \frac{1}{8} \cdot pgH^2$$

However, there is considerable variation in the dimensions of waves reaching the shore, and the difficulty of estimating the 'typical' wave height can be overcome by using the concept of the significant wave ($H_{1/3}$). This is based on estimating the height of the highest one-third of all waves observed over a period of 20 min, or the highest 33 in a train of 99 waves. The dimensions thus obtained can be used in comparisons of conditions in coastal waters at different times, or on different beach sectors.

Breaking waves are said to plunge when the wave front curves over and descends with a crash, or to spill when the crest collapses more gradually. Spilling breakers are sometimes categorised as surging where there is a strong onshore flow at the time of breaking, and subsiding where the surf declines in front of the breaking wave, but in relation to beach morphology the important point is whether the swash and backwash generated by breakers achieves net shoreward or seaward movement of beach material. In general spilling breakers have strong swash followed by a gentle backwash, producing net shoreward sediment movement, whereas plunging breakers have a short swash and a relatively stronger backwash, withdrawing sediment from the beach. This is why the former have sometimes been classified 'constructive' waves and the latter 'destructive'.

According to Johnson (1956) waves which have steepness (H_o:L_o) in deep water of less than 0.025 produce constructive breakers, while those with higher ratios become destructive. However, grain size and beach permeability also influence swash–backwash conditions: breaking waves are more likely to build up the profile on a shingle beach, where backwash energy is diminished by percolation, than on a sandy beach, where the backwash is more effective in moving sediment back into the sea. It is not uncommon for plunging breakers to pile up and steepen an upper beach of shingle or coarse sand while combing down a lower, flatter beach of finer material, a form of onshore–offshore sorting that leaves a sharp boundary between the two (Fig. 9).

Figure 9 An upper beach of shingle and a lower beach of sand on the east coast of Dungeness, south-east England, separated by swash–backwash sorting

Beach morphodynamics

Beach morphodynamics (the study of beach morphology in terms of process and change) can be interpreted in terms of beach states corresponding with surf scale categories based on the dimensions of breaking waves (Wright and Short 1984), a classification which acknowledges the interactive nature of nearshore processes and beach morphology. Dean (1973) proposed a surf scaling parameter (Ω), based on breaker height (B_h), and obtained using the formula

$$\Omega = \frac{H_b}{w_s T}$$

in which w_s is mean sediment fall velocity and T the wave period. This enables beaches to be described as reflective ($\Omega < 1$) where they receive surging breakers and have a high proportion of wave energy reflected from the beach face, or dissipative ($\Omega > 5$) where wave energy from spilling breakers is lost across a wide gently sloping beach. An intermediate category (Ω = 1 to 5) is recognised. The characteristics and dynamics of reflective, intermediate and dissipative beaches will be discussed below (p. 51).

Rip currents

Backwash is the seaward return of water that has been carried shoreward by breaking waves, but the augmentation (set-up) of nearshore sea level by shoreward movement of water due to wave motion must also return seaward, either as undertow (sheet flow near the sea bed) or (more often) in localised rip currents. Within these, water flows back through the breaker line in sectors up to 30 m wide, attaining velocities of up to 8 km/h before dispersing seaward. The shoreward movement of breaking waves and the seaward return currents form the main driving components of the nearshore water circulation.

Rip currents occur in distinct (though variable) patterns on many beaches. A light or moderate swell produces numerous small rip currents, and a heavy swell produces a few more widely spaced and concentrated rips, fed by stronger lateral currents in the surf zone. Rip currents cut channels seaward through the nearshore zone (across any bars), and deposit fans of sediment at their seaward limits; in some cases these channels extend headward to form re-entrants in the beach (Dolan 1971). When waves arrive at an angle to the beach the rip currents head away diagonally through the surf instead of straight out to sea, and cut oblique channels through the nearshore zone. Rising tides and onshore winds raise the water level along a beach and thus intensify rip currents.

Other nearshore currents

Waves that break parallel to the beach produce orthogonal swash and backwash and mainly onshore–offshore movements of water and sediment, but there are usually lateral variations in breaker height, particularly where wave refraction has generated contrasts indicated by varying orthogonal spacing, as mentioned above. Any such lateral variations in breaker energy are resolved by divergent longshore current flow in the nearshore zone. As a rule the natural variability of wave dimensions results in energy changes that result in pulsations in longshore and onshore–offshore current flow (Komar 1976).

Also associated with waves breaking parallel to the beach are orthogonal edge waves, standing oscillations developed at right angles to the shoreline (Guza and Inman 1975, Hardisty 1994). The crests of these edge waves augment incoming breakers to form swash salients, and produce longshore flow into intervening troughs, where outflowing rip currents may begin. As edge waves have the same periodicity as incident waves the two motions combine to produce a regularity in nearshore water circulation which may explain the formation of rhythmic features such as beach cusps (Fig. 10). These are patterns of coarser spurs and finer bays that develop along the beach face in response to circulations derived from waves breaking parallel

Figure 10 Concave profile produced on a shingle beach at Ringstead, Dorset, England, cut back by preceding storm wave action. Waves have formed beach cusps on the concave slope

to the shore, their spacing being proportional to breaker height. Once formed, they also define sustaining swash and backwash patterns.

When waves arrive at an angle to the beach they deflect the nearshore water circulation, generating currents along the shore. The effects of these wave-induced longshore currents are difficult to separate from the associated effects of oblique swash and orthogonal backwash produced when waves arrive at an angle to the shore and break on the beach, for both processes move sediment alongshore, the action of oblique wave swash causing lateral beach drifting, while the wave-induced currents cause longshore drifting in the nearshore zone (Fig. 11).

Onshore–offshore current movements accompany breaking waves and the ensuing swash and backwash, and include rip currents, as previously described. Rising and falling tides generate weak shoreward and seaward currents, and when winds blow from land to sea they can set up a shoreward bottom current in the nearshore zone, which may move sediment on to a beach (King 1972).

Dominant waves

The direction, height and periodicity of waves approaching a beach have a strong influence on its plan and profile, and when wave conditions change

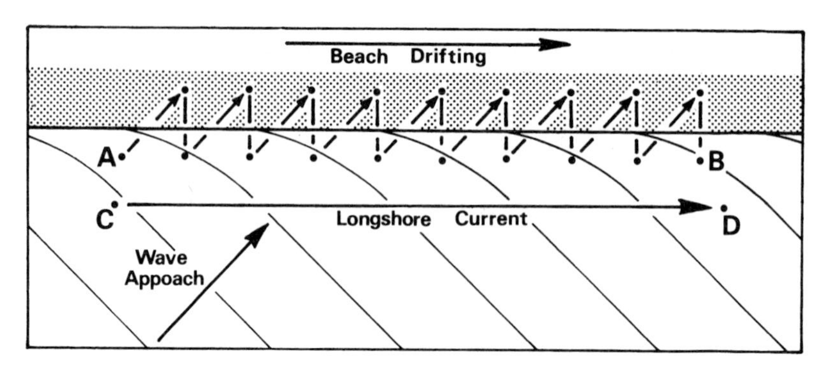

Figure 11 Longshore drifting is the movement of sediment which occurs when waves arrive at an angle to the coast. It is the result of beach drifting produced by swash and backwash (A→B) and the associated longshore current (C→D) generated in nearshore waters

there are corresponding adjustments in beach morphology. Wave conditions change frequently, and it is necessary to analyse several years' records to establish characteristic annual and seasonal wave regimes. Direct observations of wave conditions are made from lightships and certain coastal stations, and these can be supplemented by analyses of meteorological data, when the direction and strength of winds are correlated with the pattern and dimensions of locally generated waves. Dominant waves tend to determine such features as beach alignment and the net direction of longshore drifting and sorting on beaches. Variations in coastal aspect can result in contrasts in longshore drifting in response to dominant waves, as shown in Fig. 12.

On many coasts a particular wave direction is clearly dominant: the south coast of England is dominated by waves from the direction of the prevailing south-westerly winds, and much of the east coast of Australia has a prevalence of south-easterly wave action. Southerly and south-westerly swell is dominant on many oceanic coasts (p. 22), including the Pacific coasts of the Americas, the Atlantic coast of Africa, the coasts of the Indian Ocean from Arabia round to Western Australia, the south coast of Australia and the west coast of New Zealand. Other coasts show greater variation: the North Norfolk coast in England, for example, has waves arriving from the north-west, north and north-east according to local wind conditions, which change with the passage of depressions and anticyclones. In some years the north-easterly waves are dominant, in others north-westerly waves prevail. The east coast of Port Phillip Bay, Australia, has seasonal variations in dominant waves, with westerly and south-westerly waves prevalent in summer and north-westerly waves commoner in the winter months. Correlation with meteorological patterns can be used to determine long-term dominant wave directions in such conditions.

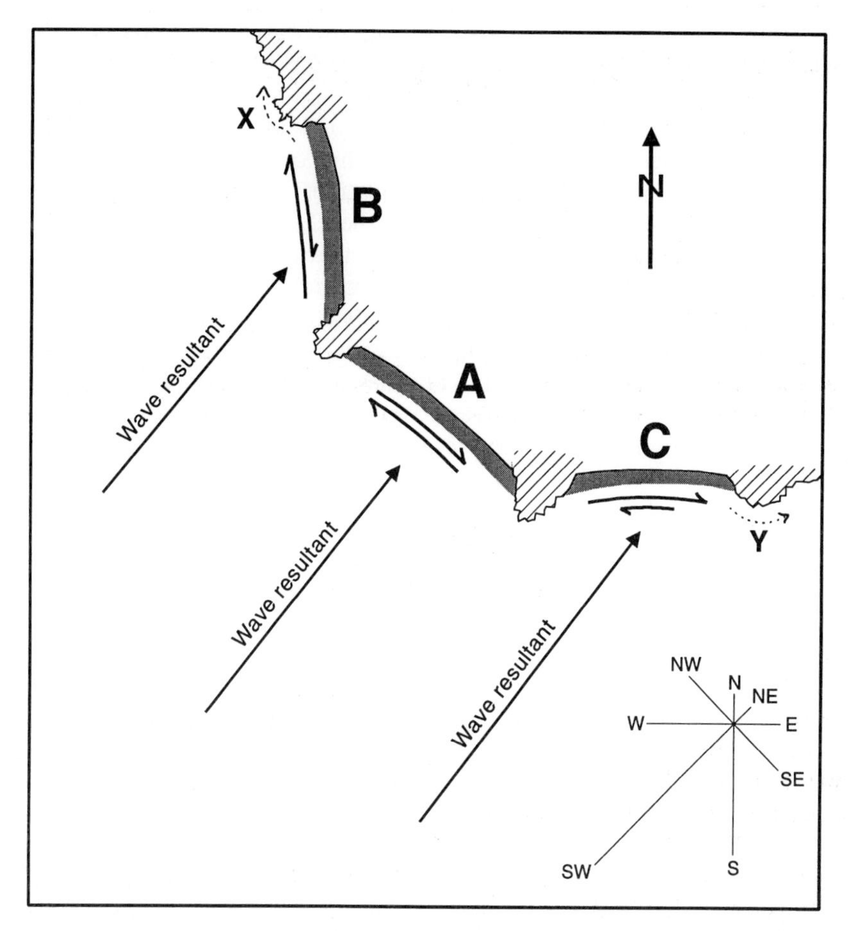

Figure 12 The influence of aspect on longshore drifting. On beach A, with an orthogonal wave resultant, there is no net drifting; on beach B northward drifting is dominant, and on beach C eastward drifting is dominant. Losses are likely round headlands at X and Y

High, moderate and low wave energy coasts

Coasts exposed to ocean swell and stormy seas are known as high wave energy coasts, while those that are sheltered from strong wave action, bordering narrow straits, landlocked embayments, or island or reef-fringed seas with limited fetch, or where wave energy has been reduced by intense refraction, are termed low wave energy coasts. It is useful to recognise an intermediate category of moderate wave energy coasts.

High wave energy coasts can be defined as those with mean annual significant wave height (measured as the waves break) exceeding 2 m, moderate wave energy coasts 1 to 2 m, and low wave energy coasts < 1 m. Beaches on high wave energy coasts, as on the stormy Atlantic shores of Europe, and the ocean coasts of southernmost Africa and southern and western Australia have bold, sweeping outlines. These become more irregular on moderate wave energy coasts, while on low wave energy coasts beaches are typically shorter, with such features as cusps, lobes and spits along the shore. Examples are found in the Danish archipelago, around Puget Sound, and on the shores of many estuaries and lagoons. The Gulf Coast of Florida, in the United States, has generally low wave energy, wave action being reduced by the broad, gently-shelving offshore profile. It shows an intricate shore configuration with only limited beaches interspersed with minor spits, deltas and marshes, and persistent offshore shoals. However, this generally low wave energy coast is occasionally subject to drastic modification by strong wave action during hurricanes. Beaches are then overwashed, and the coast may develop new features that persist during subsequent calmer conditions.

As has been noted, grain size of beach sediments is not necessarily well correlated with wave energy sectors. Storm-piled cobble beaches are more likely to be found on high wave energy coasts, but many of these have sandy beaches. On the other hand beaches on low wave energy coasts may be gravelly, even though wave action is rarely strong enough to mobilise the coarser material.

TIDES

Tides are movements of the oceans set up by the gravitational effects of the sun and the moon in relation to the earth. The ebb and flow of tides produces regular changes in the level of the sea along the coast, and generates tidal currents. The rise and fall of the tide on a coast is measured by tide gauges, located chiefly at ports. The tides recorded shortly after new moon and full moon, when earth, sun and moon are in alignment and have combined gravitational effects, are relatively large, and are known as spring tides. Maximum spring tide ranges occur about the equinoxes (late March and late September), when the sun is overhead at the equator. At half-moon (first and last quarter) the sun and moon are at right angles in relation to the earth, so that their gravitational effects are not combined; tide ranges recorded shortly after this are reduced, and are known as neap tides. Stronger tidal currents are generated by spring tides than neap tides.

Tide range in mid-ocean is small, of the order of 0.5 m, but it increases where the tide invades shallow coastal waters, particularly in gulfs and embayments. Where the spring tide range is less than 2 m the environment is termed microtidal, between 2 and 4 m it is mesotidal, and if more than

4 m is macrotidal (Davies 1980). Microtidal ranges are typical of the coasts of the Atlantic, Pacific, Indian and Southern Oceans and of certain landlocked seas (Baltic, Mediterranean, Black, Red and Caribbean). Around the Pacific Basin mean spring tide range exceeds 2 m only in certain gulfs, notably the Gulf of Siam (Bangkok, 2.4 m) and the Gulf of California (9 m). On the Australian coast microtidal conditions prevail from Brisbane south to Cape Howe, then along the whole of the southern and western coast, with the exception of certain embayments. Tide range increases into Spencer Gulf (Port Augusta, 2.0 m) and Gulf St. Vincent (Port Wakefield, 2.1 m), but the narrow entrance to Port Phillip Bay (see Fig. 76, p. 144) impedes tidal invasion, so that tide ranges within the bay are only about half those recorded on the outer coast.

Macrotidal ranges are found in the Bay of St. Michel in north-west France, where maximum tides exceed 9 m, the Bristol Channel, where mean spring tide range increases eastwards to attain 12.3 m at Avonmouth, and the head of the Bay of Fundy, where tides of more than 15.2 m are recorded. It is thought that the world's largest tide ranges, more than 20 m, occur within Leaf Bay, in north-east Canada. There are macrotidal coasts in northern Australia, especially between Port Hedland (up to 5.8 m) and Darwin (up to 5.5 m), and very large ranges are recorded in the gulfs, Derby, on the shore of King Sound, having a spring tide range of 10 m.

A large tide range implies a broad inter-tidal zone, more than 20 km of sandflats and mudflats being exposed at low spring tides in the Bay of St. Michel. Wave energy is expended in traversing such a broad shore zone, and waves which reach the beaches at high tide have been much diminished by friction in the shallow water passage.

The influence of tides on beaches depends largely on tide range, which determines the zone over which wave action can operate. On macrotidal coasts the beach is usually found at about high tide level, and is washed by waves only briefly and intermittently, wave action being withdrawn from the backshore for most of the tidal cycle. Re-shaping of the beach profile, sorting of beach sediments and modification by longshore drifting are therefore all retarded by the dispersal of wave energy as the tide rises and falls, while tidal currents play an important part in the shaping of the topography exposed at low tide: a topography which can be termed tide-dominated. Strong tidal currents that develop in inlets and constricted bay mouths can scour bordering beaches and interrupt longshore drifting of beach sediments, sometimes curtailing spit growth (e.g. Sandy Point, Fig. 13).

On microtidal coasts, by contrast, wave energy is concentrated within a narrow vertical zone, and the beach is more consistently subject to wave processes. Beach morphology, sediment transport and nearshore sea floor topography all become wave-dominated in these situations.

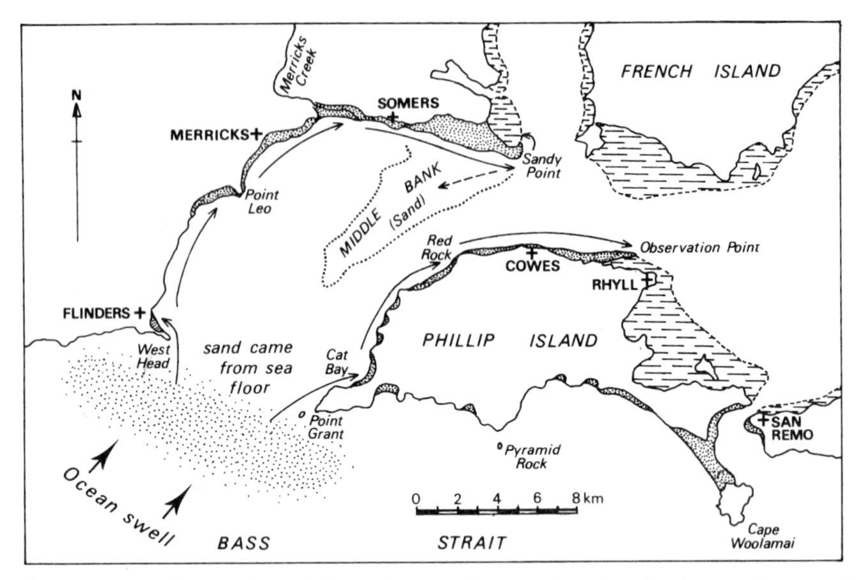

Figure 13 Delivery of sand from the sea floor to beaches in the south-west of Westernport Bay, Australia. Drift-dominated beaches have formed as longshore drift carried sand along the coasts from Point Leo to Sandy Point and from Cat Bay to Observation Point

SHORT-TERM SEA LEVEL VARIATIONS

Sea level also shows short-term variations resulting from meteorological effects. Onshore winds build up the level of coastal waters, whereas offshore winds lower them. A fall in atmospheric pressure is accompanied by elevation, and a rise by depression, of the sea surface: a fall of 1 millibar raises the water level about a centimetre. Beaches respond to such short-term variations in much the same way as to the more regular tidal oscillations.

Storm surges occur when strong onshore winds are produced by deep depressions (cyclones, hurricanes and typhoons in the tropics). These build up coastal water to an exceptionally high level for a few hours or days, and are most pronounced when they coincide with high spring tides. The strong onshore winds also generate large waves to accompany the raised sea level, overwashing beaches and low-lying coastal areas, and causing extensive changes in a short period. Beach erosion is usually severe during a storm surge, and if the coast consists of soft formations it may be cut back substantially to form a new morphology. One of the best documented storm surges occurred in the North Sea in 1953, when a northerly gale raised the sea level up to 3 m, flooding extensive areas in eastern England (Grove 1953) and low-lying parts of the Dutch and German coasts. A similar surge accompanied the passage of a deep depression across the North Sea in 1978, and there have been many such episodes in this region in recent centuries, when beaches have been eroded and briefly submerged.

Hurricanes raise sea level temporarily by as much as 6 m, causing extensive erosion and damage to structures along the shores of the Atlantic and Gulf coasts of the United States, and major storm surges occur from time to time at the head of the Bay of Bengal, when large areas of the Sundarbans are devastated by submergence. In northern Australia tropical cyclones generate surges in a region where heavy rains accompany the summer monsoon, and cause widespread flooding in coastal districts. Beach erosion is severe during such episodes, and some features, such as piled-up beach rock (p. 39) or overwashed fans of beach sediment, may persist for years or decades after such catastrophic events.

OTHER SEA DISTURBANCES

Apart from storm surges, exceptional disturbances of ocean water occur during and after earthquakes, landslides or volcanic eruptions in and around the ocean basins. These produce giant 'tidal waves' or tsunamis, which may attain heights of more than 30 m by the time they reach the coast. They are most common in the Pacific Ocean, which is bordered by zones of crustal instability, and they are responsible for occasional catastrophic flooding and beach erosion on Pacific coasts, as in 1946, when a tsunami was initiated by an earthquake off the Aleutian Islands, and waves travelling southward caused extensive devastation in the Hawaiian Islands, 3700 km away.

In 1883 the explosive eruption of Krakatau, a volcanic island in Sunda Strait, Indonesia, generated a tsunami up to 30 m high on the nearby coasts of Java and Sumatra, sweeping away beaches and hurling coral boulders and blocks up on to fringing reefs and shore platforms.

Huge waves can be generated in restricted areas by landslides and rock falls. A massive rock fall into Lituya Bay, an Alaskan fiord, in 1960 swept a wave 15 m high down the fiord at 160 km/h, washing over a spit at the entrance and dispersing beach sediments. Disturbances of a similar kind occur in the vicinity of ice coasts as the result of iceberg calving.

These are unusual phenomena. On most coasts waves influence the shaping of beaches within a zone above and below present mean sea level, the extent of this zone depending largely on tide range. Rapid changes occur on beaches during storms and occasional storm surges, but there are also more gradual gains and losses, leading to the re-shaping of the beach in plan and profile during long intervening periods of less boisterous weather.

CURRENTS

Reference has already been made to currents associated with wave action and currents generated by tides. Other currents include ocean currents, gentle movements of water in response to prevailing wind patterns, and density variations in the oceans resulting from differences in the salinity and temperature of the water. They have little effect on beaches or nearshore morphology, except where they bring in warmer or colder water, which modifies ecological conditions and thereby influences the distribution of such features as coral reefs or kelp beds, the presence of which can affect beach forms and dynamics.

Tidal currents, produced as tides rise and fall, alternate in direction in coastal waters, reversing as the tide ebbs; their effects may thus be temporary or cyclic. In the open ocean tidal currents rarely exceed 3 km/h, but where the flow is channelled through gulfs, straits between islands, or entrances to estuaries and lagoons, these currents are strengthened, and may locally and briefly exceed 16 km/h. The strongest currents are generated in macrotidal areas at spring tides. Tidal oscillations impinging on a coastline may set up longshore currents, as on the North Norfolk coast in England, where the longshore flow is westward during the two to three hours preceding high tide, then eastward for another two to three hours as the ebb sets in. These currents are of limited importance in terms of beach erosion or deposition, but they carry sediment along the coast in the nearshore zone, and this may eventually be delivered to beaches alongshore. The effects of tidal currents are otherwise generally subordinate to the effects of waves and associated currents in the nearshore zone. These associated currents include wind-driven currents produced where winds move the surface water, building up sea level to leeward and lowering it to windward, as wave action proceeds. Wind-generated currents are not as regular as the alternating tidal currents, but their effects are likely to be cumulative in the direction of the prevailing wind. Mention has been made of the hydraulic currents (backwash, undertow, rips) that then disperse water from areas where it has been piled up, flowing until normal levels are restored.

Strong currents are produced when winds drive surface water into gulfs, through narrow straits, or in and out of estuary and lagoon entrances. These may strengthen or oppose the currents produced by tides in similar situations, and it is often difficult to separate the effects of the two.

Currents produced by discharge from river mouths carry sediment into the sea, maintain or enlarge river outlets, and form a seaward 'jet' which refracts approaching waves and can act as a breakwater impeding or interrupting longshore currents and sediment flow. Fluvial discharge currents are strongest off streams fed by melting ice and snow from coastal mountains, as in Norway and Alaska during the summer months. In tidal estuaries, fluvial discharge is reduced and possibly halted by the rising tide, but augmented by the ebb. Similar currents are produced through tidal inlets by the ebb and flow of lagoon water and the outflow from rivers that drain into the lagoon.

Currents can readily move fine to medium sand (grain diameters between 0.1 and 0.5 mm) when their velocity exceeds about 15 cm/s, but stronger currents are required to move coarser material. Currents generated by winds and tides can be strong enough to move sand or even gravel on the sea floor, either contributing to longshore drifting of beach sediment, supplying material to a beach, or carrying it away offshore, but these effects are usually subordinate to the onshore–offshore and alongshore movements of beach sediment by wave action. Currents can prevent nearshore deposition and erode channels or remove shoals, thereby deepening the nearshore water and permitting larger waves to break upon a beach. Sea floor shoals of sand or gravel may form adjacent to rocky headlands as the result of deposition by current flow, examples being the Skerries, off Start Point, and the Shambles, off Portland Bill, on the south coast of England. Material from such shoals may in due course be delivered to a beach, but the more immediate effect is to modify wave refraction patterns in nearshore waters and reduce wave energy reaching the nearby shore.

In general, currents are more effective in shaping sea floor topography than in developing beach configuration, and the early theory that long, gently curving beaches on oceanic coasts were produced by currents sweeping along the shore has given place to the modern view that these outlines are determined by refracted wave patterns. Nevertheless, changes in the topography of the sea floor, due to erosion by current scour or deposition from slackening currents, modify patterns of wave refraction and may thus indirectly affect beach outlines. Currents often play a part in removing material eroded by waves from the coast, or in supplying the sediment that is subsequently built into beaches by wave action.

NEARSHORE WATER CIRCULATION

The combined effects of wind-generated waves, astronomically generated tides, various forms of current flow, and other disturbances of the sea produce a highly variable energy flux in nearshore waters. As has been noted, the various processes interact: a rising tide, for example, deepens the nearshore water (the nearshore zone being generally concave upward in profile), thereby increasing the height and energy of waves that reach the shore. A tidal current flowing in one direction can reduce the velocity and dimensions of waves moving in a contrary direction. Current flow in the nearshore zone is the resultant of potential flows generated by winds, waves, tides and other forces, and there is much variation in current direction and velocity. In addition, wave variability results from the arrival of waves generated from differing distances and directions, and there are often intersecting wave patterns arriving in the nearshore zone. The outcome is a complex nearshore hydrodynamic system (certainly different, and possibly more complex, than river hydrodynamics) that influences the shaping of the coastline, including beaches, and the nearshore sea floor.

Moreover, studies of the effects of nearshore processes on beach morphology and sediments have made it clear that there is an important feedback component, whereby coastal configuration, beach morphology and nearshore features influence the processes at work on them in the nearshore zone (Krumbein 1963, Komar 1976). The attempt to explain beach features and changes as responses to the nearshore process regime should take account of this feedback, acknowledging that there is a tendency towards an adjustment between nearshore processes and beach morphology. Nevertheless, beaches are shaped mainly by breaking waves, and to a lesser extent by the effects of nearshore currents. Numerous studies of beach systems have shown how nearshore processes influence both the plan and the profile of beaches, a good recent example being the analysis by Short (1992) of beach systems on the central Netherlands coast.

Most beaches are changing almost continuously in response to rising and falling tides and to variations in waves, currents and wind action related to weather conditions. Other factors influencing their shaping will now be considered in more detail.

WEATHERING AND INDURATION OF BEACH MATERIAL

As beach sediment is agitated by wave action, sand and gravel particles are worn into rounded shapes and also reduced in size by attrition. The rate of such physical weathering depends on the hardness and structure of the grains: quartz sand and flint gravel are relatively durable, whereas sandstone fragments may disintegrate and disperse (an example is discussed on p. 68). Weathering thus contributes to the selective sorting of a beach sediment.

As rounding and attrition proceed, sand grains become smooth and highly polished, while pebbles and cobbles tend to become slightly flattened, and thinner at right-angles to the longest axis. Beach volumes are also modified by chemical weathering, notably the dissolving and removal of carbonates (shells, coral or limestone fragments) by percolating rain water, corrosive groundwater, and sea water. Usually sea water is saturated with dissolved carbonates, but at lower temperatures (at night, during winter, or in cold climates), and in aerated sea spray, the carbon dioxide content rises, causing acidification, which leads to further solution of carbonate rocks and sediments.

Conversely, the precipitation of carbonates in the zone of fluctuating water table within a beach (related to the rise and fall of tides and alternations of wet and dry weather) can cement beach sand into hard sandstone layers known as beach rock, which may be exposed by subsequent erosion. Beach rock is frequently exposed on tropical beaches, especially coral cays, but it can occur on temperate coasts where cementing carbonates are provided by seepage from the hinterland. Where the cemented beach material includes angular gravel it is termed a beach breccia, and where rounded pebbles are enclosed, a beach conglomerate.

Another form of induration is seen where a beach surface develops a coherent crust as the result of interstitial accumulation of fine-grained sediment, which binds the sand or gravel, as the result of surface compaction (p. 103), or as the result of cementation by precipitated carbonates. Where such a crust has developed, wave action on the beach face may cut a small cliff, capped by the indurated layer: a process known as scarping (p. 158).

Longshore drifting

Apart from the effects of weathering, beach sediments are washed by wave action into a variety of forms as they accumulate or as they are eroded (Hayes 1972). Movement of beach material along the coast occurs when sand and shingle are edged along the shore by waves that arrive at an angle to the shore and produce a transverse swash, running diagonally up the beach, followed by a backwash that retreats directly down into the sea (Fig. 11). Such beach drifting is accompanied by sediment flow in the nearshore zone, generated by the longshore currents that accompany such obliquely arriving waves. The combined effect of these processes is longshore drifting of sediment to beaches and spits downdrift. Longshore drifting is most rapid when wave crests approach the shoreline at an angle of between 40° and 50°, where the shoreline is straight or gently curved and unbroken by headlands, inlets or estuaries, and where the nearshore sea floor profile is smooth.

Beach sediments may be moved first one way, then the other, according to the direction from which the waves approach. If waves arrive as frequently from one direction as the other, the resultant drift over a period will be negligible, but there is usually a definite long-term drift in one direction, indicated by the longshore growth of spits, the deflection of river mouths, or the accumulation of beach sediment alongside headlands or breakwaters (Jacobsen and Schwartz 1981). Thus on the north Queensland coast in Australia, the dominant waves generated by the prevailing south-easterly trade winds have caused longshore drifting of sand northward from river mouths to form a succession of beaches and spits that deflect the river mouths northward. The Burdekin River, for example, has a delta with sandy beaches extending northward from the mouths of each of its distributary channels, culminating in the long recurved Bowling Green spit. Accretion on the southern side of the harbour breakwater at Mackay is another indication of this northward drifting.

Beaches showing alternations of longshore drifting as waves arrive from each direction occur on the Pacific coast of North America because of seasonal changes in the direction of incident swell (p. 22). In Half Moon Bay, California, waves from the south-west move sand northward in summer and waves from the north-west drive it back to the south in winter (Emery 1960). Further north on the Pacific coast, the Columbia River has delivered large quantities of sand which have drifted northward in summer when the waves arrive mainly from the south-west, and southward in winter when north-westerly waves prevail. The stronger northward drifting has carried sand up the Washington coast to supply beaches as far as Cape Shoalwater, and the southward drifting has built the beach that extends down to Tillamook Head. On the north-east coast of Port Phillip Bay, Australia, beaches diminish at the southern end and increase at the northern end during summer, with a reversal of this pattern during winter, in response to seasonal variations in the direction of onshore wind and locally generated wave action (Fig. 14) (Bird 1993a).

Other features produced by longshore drifting are discussed below (p. 57).

BEACH COMPARTMENTS

Many beaches occupy distinct compartments, bounded by rocky reefs or protruding headlands, particularly those that end in deep water (Davies 1974). There can be striking differences in the nature of beaches in adjacent bays, as on the steep coast of south-west England, where the grain size or mineral composition of beaches varies from one cove to another, the beach compartments being separated by rocky promontories. Within each compartment wave action can move the confined beach material to and fro along the shore, such alternations of longshore drifting building up first one side then the other, or change the transverse profile by alternations of shoreward and seaward movements of sediment.

If the volumes of sediment gained and lost on each sector within a beach compartment over particular periods are calculated, these changes can be expressed in terms of a beach budget for that compartment. Beach budgets can also be established for a coastline that includes more than one beach compartment, or for any other defined sector, such as a seaside resort waterfront. Figure 15 shows an example, based on regular surveys of profiles at intervals along a beach near Somers, in Westernport Bay, south-eastern Australia: volumes of sand in arbitrary beach compartments

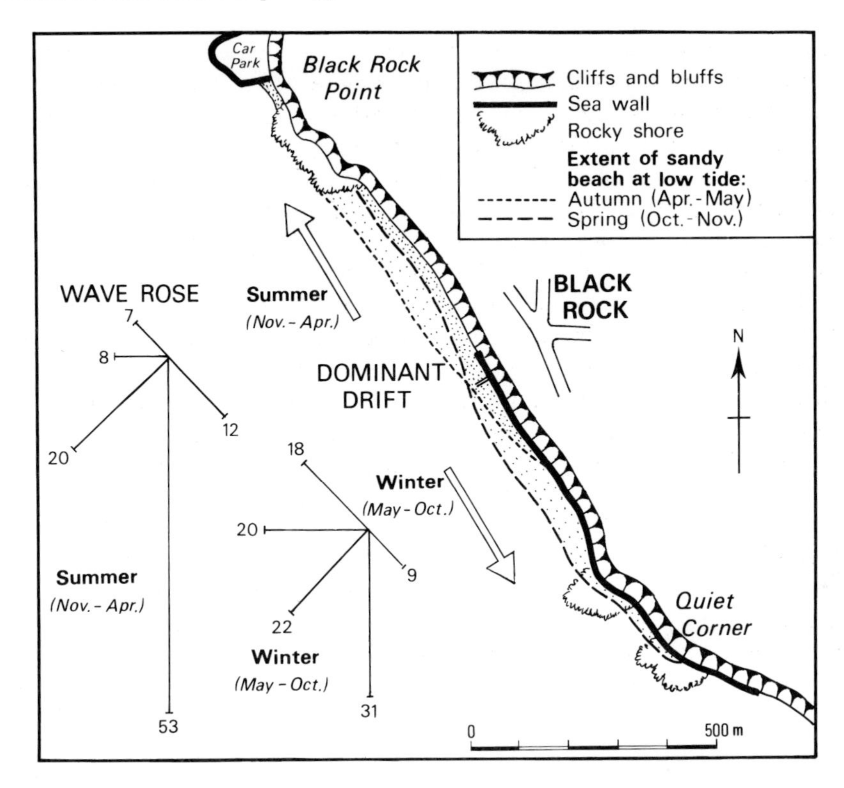

Figure 14 Black Rock Beach, where northward drifting is dominant in the summer (November to April) and southward drifting in winter (May to October). The wave roses are based on the percentage frequency of onshore winds > Beaufort Scale 3, which determine wave action on this part of the coast of Port Phillip Bay, Australia

between successive pairs of profiles were calculated by multiplying their cross-sectional areas by the intervening distance (Bird 1985b).

There are gains and losses from beach compartments if the intervening headlands are small enough for sand and gravel to be carried past them on the sea floor by waves and currents. Such natural bypassing often takes place predominantly in one direction as a response to prevailing longshore wave and current action. Some beach compartments are delimited by river mouths or tidal entrances, where transverse ebb and flow currents impede longshore drifting in much the same way as headlands or solid breakwaters. This can result in beach accretion on the updrift side, and erosion down-drift, but waves and associated currents may carry sand and gravel across the sea floor, bypassing the mouths of rivers and tidal entrances, sometimes with the formation of longshore bars.

There are beach compartments in southern California where sand delivered to the shore by rivers drifts southward along the coast, past headlands

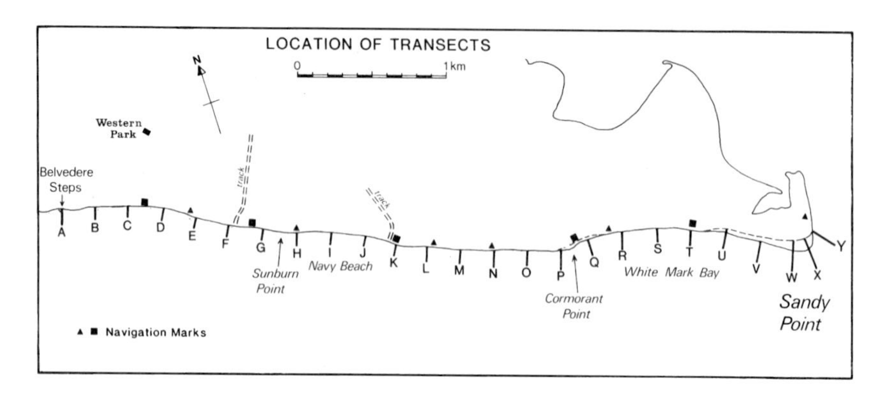

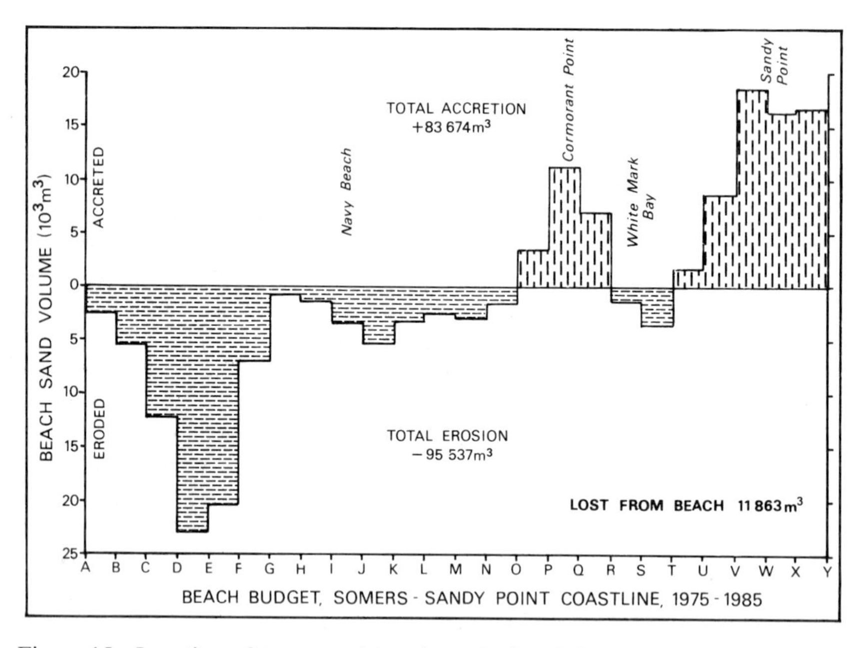

Figure 15 Location of transects (above) on the beach between Somers (Belvedere Steps) and Sandy Point, Westernport Bay. The pecked line indicates sectors of the 1975 coastline where substantial progradation occurred by 1985. The lower diagram gives a beach budget for 1975–1985, based on surveys of cross-sectional areas over the decade, multiplied by the intervening distances (generally 225 m). The volume of sand gained on accreting sectors was less than that lost from eroding sectors (Bird 1985b)

and through successive bays, until it is lost into submarine canyons that run out from the southern ends of embayments (Emery 1960). A similar situation exists on fluvially fed gravel beaches in Georgia, on the Caucasian Black Sea coast (Zenkovich 1973). The 'river of sand' model made an attractive colour movie, produced by Encyclopaedia Britannica Education Films in 1965, but the analogy between beaches and rivers is misleading because most of the world's beaches are not nourished by sand input at one end with longshore drifting to lose sediment into submarine canyons at the other. Most beaches receive sediment from several sources, including eroding cliffs and the sea floor, or from rivers supplying mid-beach situations, and very few lose sediment into submarine canyons (Tanner 1987).

BEACH OUTLINES IN PLAN

Beaches often show smoothly curved outlines in plan, concave seaward (see Fig. 8), determined by the patterns of approaching waves, so that they anticipate, and on breaking fit, the plan of the beach.

Changes in configuration occur on beaches that have not yet become adjusted to the prevailing wave regime. For example, where the beach outline is more sharply curved than the approaching swell, the breaking waves produce a convergence of longshore drifting of beach sediment towards the centre of the bay, where the beach progrades until it fits the outline of the arriving swell. Convergent longshore drifting in Byobugaura Bay, near Tokyo, has prograded the central sector at Katakai in this way, with sediment derived from the erosion of the northern and southern parts, and a similar sequence has been documented from Guilianova on the east coast of Italy (Zunica 1976). Where the shoreline is less sharply curved than the approaching swell, the waves set up a divergence of longshore drifting from the centre until the beach outline fits the wave pattern. This has occurred on the shores of the Andalusian Bight, in southern Spain, where erosion of the central sector has been balanced by progradation at Matalascanas to the south and Mazagon to the north.

Beaches shaped by waves generated by local winds (i.e. not receiving ocean swell) have outlines related partly to the direction, strength and frequency of onshore winds, partly to variations in the length of fetch (open water across which waves are generated by those winds), and partly to wave refraction. The orientation of such beaches is the outcome of the long-term effects of the waves arriving from various directions. These can be expressed as a resultant, calculated from records of the frequency and strength of onshore winds, ignoring those of Beaufort Scale less than 4 (<21 km/h) which produce only small waves that have little effect on beaches. Typically such beaches are modified by erosion and accretion until they become orientated at right angles to this onshore wind resultant.

Where there are marked contrasts in fetch, however, these must be taken into account, for strong winds blowing over a short fetch may be less effective in generating wave action than gentler winds blowing over a long fetch. If the direction of longest fetch coincides with the onshore wind resultant, the beach becomes orientated at right-angles to this coincident line, but where the two differ the orientation becomes perpendicular to a line that lies between them. On coasts of intricate configuration (as in the Danish archipelago) wave action is determined more by the fetch than the onshore wind resultant, and the beaches become aligned at right-angles to the maximum fetch (Schou 1952). The same is true of beaches on the shores of landlocked embayments or coastal lagoons.

Where waves are refracted by nearshore islands or local irregularities on the sea floor they move into the shore to form cuspate beach outlines in the lee of offshore reefs or shoals, and embayed sectors behind sea-bed hollows. If these offshore features remain in position, such cusps and bays persist in position, but changes in sea-floor configuration when currents scour out hollows or build up shoals, or when banks or shoals migrate shoreward, seaward or along the coast, modify patterns of wave refraction and so change the outlines of the beach. Thus as a shoal moves along the coast the cuspate beach in its lee is eroded on one side and accreted on the other, so that the cusp also migrates. It has been suggested that the longshore movement of sand and shingle forelands at Benacre and Winterton Ness, on the East Anglian coast, is the outcome of changing wave and current patterns related to the migration of shoals offshore (Robinson 1966).

A distinction can be made between beaches which have swash alignments (swash-dominated beaches), having been built parallel to incoming wave crests with little or no longshore sediment movement, and beaches with drift alignments (drift-dominated beaches), parallel to the line of maximum longshore sediment flow, generated by obliquely incident waves. Beaches exposed to refracted ocean swell are typically built on swash alignments, as are bay-head beaches (see Fig. 8), but waves moving into embayments often build beaches on drift alignments, sometimes terminating in spits (see Fig. 12). In general, beaches with swash alignments are smoother in outline than those with drift alignments, which are typically sinuous, with intermittent lobes and slightly divergent longshore spits and bars (Davies 1980).

The shaping of beach outlines also depends on patterns of sediment supply, with a tendency for beaches to become narrower downdrift from a sediment source such as a river mouth or eroding cliff. Beach accretion also depends on the pattern of incoming waves in relation to the source of the material. On a deltaic coast where sand or gravel are being delivered to the mouth of a river, waves arriving parallel to the coastline diverge along both sides of the delta, and distribute the fluvially supplied sediment to pro- ducing symmetrical beaches, as on the Tagliamento delta on the northern shore of the Adriatic Sea, or trailing lateral spits, as on the shores of the Ebro delta in Spain.

There is a notion that beach outlines in plan become stable (or attain an equilibrium) once they have attained a shape that fits the pattern of dominant refracted wave. The crenulate or 'half-heart' shaped beaches produced where incoming waves are refracted round an adjacent headland, as on the New South Wales coast in Australia, have been cited as an example (Silvester 1970). The inference is that coastal engineers should endeavour to create beaches of such a shape to achieve stability.

Equilibrium is defined in the Oxford English Dictionary as 'a condition of balance between opposing forces, the forces being so arranged that their resultant is zero'. A cyclic equilibrium is one that returns to its original condition after being disturbed, and a shifting or dynamic equilibrium is one that changes while remaining in balance with driving forces.

There is no doubt that once beach outlines have become adjusted to the prevailing pattern of refracted waves they are more stable than those that have not yet achieved such an adjustment, but such beaches continue to be cut back parallel to the existing coastline if they are losing sediment, or to prograde evenly if there is a continuing supply of sediment, from alongshore or from the sea floor. Stability can only be attained where there is a sufficient input of sediment to balance episodic losses, and so maintain an equilibrium in plan.

Beaches that show lateral alternations, one end being prograded as it receives sediment drifting from the other, in response to changing directions of wave approach (p. 62) can be thought of as showing a cyclic equilibrium if they return regularly to earlier configurations, providing there is no net gain or loss of beach material.

It has been suggested that a dynamic equilibrium may be achieved within a beach compartment where the beach attains a curvature adjusted in such a way that waves impinging on the shore provide precisely the energy required to transport the sediment arriving at one end of the beach through to the other, the configuration being maintained. Komar (1976, p. 241) suggested that 'with established sediment sources and under a certain wave climate, a beach will tend towards a natural equilibrium [in plan] where the waves are just capable of redistributing the sand supplied from the sources'. These definitions make an analogy with the graded stream concept, adapted for beaches with longshore sediment flow and receiving sediment, but they do not cover situations where sediment is coming in from the sea floor, river mouths or eroding cliffs, where there is no sediment supply at present, or where there are losses of sediment offshore.

Equilibrium concepts have been applied to beach outlines in profile (p. 53) as well as in plan. It is possible for beach outlines in plan to change as the result of gains or losses in beach volume resulting in the advance or retreat of beach profiles on particular sectors, or for outlines in plan to persist even though gains or losses have steepened or flattened beach profiles.

BEACH PROFILES

Beach profiles are also shaped largely by wave action, notably the swash generated as waves break upon the shore and the ensuing backwash (p. 25). Swash and backwash velocities can be measured with a dynamometer in the waves close to the shore, and correlated with onshore and offshore movements of beach sediment as indicated by profile changes or accumulations in trays placed to intercept beach material drifting shoreward or seaward. As has been noted, spilling waves, with a strong swash and lesser backwash, tend to move sediment up the beach, building up a convex profile, and plunging waves, with limited swash and strong backwash, tend to withdraw it to the nearshore zone, leaving a concave profile of the kind seen in Fig. 10.

The transverse slope of a beach profile is also related to the grain size of beach sediments. Gravels tend to assume steeper gradients than sand (see Table 2), chiefly because they are more permeable. As the waves break, swash sweeps sediment up on to the beach, and backwash tends to carry it back, but the greater permeability of gravel and coarse sand beaches absorbs and diminishes the effectiveness of backwash, so that swash-piled sediment is left at relatively steep gradients. Fine sand beaches are more affected by backwash, and develop gentler slopes as more sediment is withdrawn seaward.

A beach with a mixture of sand and gravel (or fine and coarse sand) may be sorted by swash and backwash until the profile consists of a coarser upper beach and a finer lower beach, exposed as the tide ebbs (p. 26). The upper beach is often of shingle, clearly demarcated from the lower beach of sand, a contrast resulting from the differing responses of shingle and sand to storm waves, which pile up swash-built shingle that persists as backwash withdraws the sand (see Fig. 9). The withdrawn sand is sometimes deposited in the form of a step at the foot of the beach, which may be prograded seaward as the tide falls.

Sorting of sediment by wave action may also result from swash concentrating the coarser and heavier material at the back of a beach while backwash carries away the lighter sand grains. Selection of coarser sediment or heavy minerals in this way may lead to the formation of stratified (or laminated) deposits, with coarser seams between layers of finer sediment, seen when a section is cut across a beach, and recording alternations of stronger and weaker wave action during a phase of overall beach accretion. Within each stratified layer the distinctive sediment is well sorted, being related to a particular phase of deposition or reworking by waves.

Figure 16 Shingle berm built by storm waves on the beach at Redcliff Point, east of Weymouth, England. The beach has become reflective (p. 27), the wave in the foreground having broken and been partly reflected seaward

Superficial sorting often produces zones of shingle, shelly gravel and coarse and fine sand parallel to the shoreline, sometimes arranged by edge waves (p. 28) into beach cusps (see Fig. 10).

Deposition by swash can also build up and prograde a flat beach terrace at about high tide level, or form a ridge, known as a berm, parallel to the shoreline along the length of the beach (Fig. 16). On sandy beaches the force of breaking waves may compact the sand so firmly that it is possible to ride a bicycle, drive a car or land an aircraft on a beach terrace or berm (Fig. 17).

Beach profiles are also modified by wind action, when sand is blown along or across the beach, lowering some parts and building up others. Anemometers and intercepting trays can be used to measure quantities of sand moved across or along beaches in relation to the direction, strength and duration of wind action. Onshore winds carry sand to the backshore, where dunes may be formed, and these may either be swept inland or retained by vegetation behind the beach. A foredune is a ridge built parallel to the shoreline at the back of a beach, often on a berm or beach terrace, where sand accumulates in colonising vegetation, sometimes germinating from seed-bearing litter deposited at the limit of wave swash (Hesp 1984, Bird and Jones 1988). With continuing beach accretion (particularly during a phase of falling sea level) successive foredunes may be built, marking

Figure 17 Berm of fine sand built by calm weather ocean swell on the Ninety Mile Beach, Australia, which is firm enough to land an aircraft. Viewed from the crest of a grassy foredune (p. 47)

stages in shore progradation. Beach erosion may result in cliffing of the seaward margins of foredunes, or even removal of some of the younger foredune ridges. Under these circumstances, the beach regains some of the sand previously lost to backshore dunes as the result of onshore wind action, so that over time there are sequences of beach–dune interactions.

Beach erosion may also initiate blowouts in vegetated backshore dunes, wind-excavated hollows that enlarge and grow landward, with spilling noses of loose sand and trailing parallel arms, but these can also develop as the outcome of vegetation disruption by fire, grazing, trampling or traffic. Further details on the origin and evolution of coastal dunes are available in the textbooks mentioned at the beginning of this chapter, and in works by Goldsmith (1978), Short and Hesp (1982) and Nordstrom et al. (1990). Problems of dune management are considered further in Chapter 6 (p. 216).

Onshore winds supply sand to backshore dunes, but offshore winds sweep it back across the beach and into the sea, to be reworked and transported by waves and currents in the nearshore zone. Winds blowing along the beach can carry substantial amounts of sand, especially when the beach face is dry (Fig. 18), thereby augmenting the effects of longshore drifting by waves and nearshore currents (p. 39). This has been demonstrated by So (1982) on the sandy beach at Portsea, in Victoria, Australia. Fine sand is more readily mobilised and transported by wind action than coarse sand, and pebbles (lag gravel) are left behind when beaches of sand and gravel are winnowed. A strong wind can produce a sheet of drifting sand, with sand grains airborne (in suspension), bouncing (saltation) and

Figure 18 Eastward drifting of sand on the beach at Bray, north-eastern France, as the result of a strong westerly wind. The eastward movement is indicated by the pattern of dry blown sand diverging around the blockhouse on the beach

rolling along the beach. Under these conditions beach barchans may form, and migrate downwind.

Beach profiles are lowered and cut back by strong wave action, especially during stormy periods. Currents generated by winds, waves or tides move sediment in the nearshore zone (p. 39) and within the intertidal zone as the tide rises and falls. As currents move sand or gravel offshore, onshore or alongshore, they promote erosion or accretion, and so modify the nearshore profile. In this way, they influence the form of the waves that subsequently break upon the beach, and so contribute indirectly to the shaping of the beach, in plan as well as profile. Runoff and seepage during heavy rain or as the result of melting snow or ice can also modify the beach profile by washing sand or gravel into the sea.

Beach profiles change as sea level rises or falls. Short-term modifications accompany the regular rise and fall of tides, but more substantial changes occur during sea level oscillations of longer duration. In general, a phase of rising sea level leads to recession of the beach and withdrawal of sediment seaward (the Bruun Rule; p. 78), while emergence is usually accompanied by progradation with sediment moving shoreward. Changes in response to a rising or falling sea level are complicated by variations in the availability of beach sediment. A beach receiving abundant sediment (e.g. from a nearby river mouth or rapidly eroding cliffs) may prograde even during a phase of submergence, while a beach that is losing sediment offshore or alongshore may be cut back even during a phase of emergence.

The profiles of shingle (gravel) beaches are generally steeper than those of sandy beaches. This is partly due to their greater permeability, but there is also a somewhat different response to breaking storm waves. In addition to scouring shingle away from the beach face, breaking storm waves throw some of it forward, up the beach, to build a berm at the swash limit. In this way, a stormy phase leads to the steepening of the beach profile. Prograding shingle beaches are sometimes backed by parallel beach ridges, the outcome of successive storms, each of which threw up a ridge of shingle parallel to the shoreline. On the North Norfolk coast the 1953 storm surge scoured away sandy beaches on Scolt Head Island, but piled up the crests of shingle beaches, parts of which were rolled landward, as at Blakeney Point (Steers 1964); both sectors of coastline retreated as a result.

Coarse sand can also be built into berms by storm waves which erode beaches of finer sand. Storm waves accompanied by an onshore gale, especially at high tide, can cause sand to be washed and blown from the beach to the backshore. The gradient and morphology of the nearshore sea floor is a relevant factor, for where it is gentle or diversified by bars, storm waves lose some of their energy and break as constructive surf, with a strong swash, instead of the plunging breakers seen on steeper shores. Under these conditions even medium to fine sand can be built up as terraces or berms along the shore.

The profile of a beach at a particular time is determined largely by wave conditions during the preceding few days or weeks, although the effects of a severe storm may still be visible several months later. In phases of calm weather low waves produce spilling breakers with a constructive swash which moves sand or shingle shoreward to build up a beach terrace or berm. In rough weather higher and steeper waves form plunging breakers, with collapsing crests which produce less swash, but a more destructive backwash which tends to withdraw sediment from the beach, leaving an eroded profile, typically a gently inclined or concave slope backed by a cliff cut into beach sediments or backshore dunes (pp. 75, 76). Phases of stormy weather and periods of relatively calm weather thus produce alternations of removal and replacement of beach sediment, a sequence known as 'cut and fill'. Such a sequence occurs over time scales varying from a few days to several years, but as storms are usually more frequent in winter, many beaches show a seasonal sequence of winter 'cut' and summer 'fill'. On monsoonal coasts beach erosion occurs during storms in the wet season, with recovery in calmer dry season weather. The Darwin beaches in northern Australia are cut back and lowered during the wet summer and rebuilt in the drier winter (Gowlland-Lewis et al. 1996), and similar seasonal cut and fill is seen on the east coast beaches of Peninsular Malaysia, where erosion prevails during the north-east monsoon (Teh Tiong Sa 1985).

SWEEP ZONE

As a result of such alternations over varying periods of time, the beach profile shows changes in form, being higher and wider after phases of accretion and lower and flatter (with a backing rise or cliff) after phases of erosion. The vertical section of a beach subject to these alternations is known as the sweep zone. It can be measured by making successive surveys across the beach to determine the cross-sectional area between the accreted and eroded profiles, and when this is multiplied by the length of the beach the volume of sand gained or lost can be calculated. On some beaches the sediment removed during erosional phases is entirely replaced during subsequent accretion. On many beaches, however, the erosional losses have only been partly compensated by subsequent accretion, and as the beach loses sediment the shoreline retreats. Beaches with a more abundant and sustained sediment supply show the opposite, losses in erosional phases being more than compensated by intervening accretion, so that beach volume increases and the shoreline progrades.

Sand and gravel removed from a beach during a storm may be spread across the nearshore sea floor as a terrace, but are usually shaped into bars and troughs running more or less parallel to the coastline. With the formation of these features the beach profile may actually become wider at low tide after a phase of erosion.

BEACH STATES

Surf scaling parameters (p. 27) have been used to classify the interactive relationships between beach profiles and wave conditions in terms of reflective, intermediate or dissipative states. Thus on the New South Wales coast, where fine-to-medium sandy beaches occupy asymmetrical embayments with micro-tidal moderate-to-high wave energy conditions, Wright and Short (1984) made a distinction between steep beaches facing relatively deep water (without bars) close inshore, which are reflective because part of the incident wave energy (plunging waves) is reflected seaward (see Fig. 16), and flatter beaches fronted by nearshore sand bars and wide surf zones, which are dissipative because much of the wave energy (spilling waves) is lost as waves arrive through the shallow water (see Fig. 47). The surf zone, where the waves have broken, is often less than 10 m wide on reflective beaches, but becomes at least 100 m wide on dissipative beaches. Reflective beach profiles have upper slopes of between 6° and 12°, often with a steeper step at the base, then a gentler nearshore slope of 0.5° to 1°, and with well-formed ripples parallel to the shoreline. Dissipative beach profiles are broader and flatter, with slopes typically less than 1°, and a nearshore zone diversified by multiple parallel sand bars. Studies of beaches in other parts of Australia and around the world have shown an association of high, steep waves and fine sand on dissipative beaches, while long, low waves and coarser sediment (including shingle) characterise reflective beaches. An intermediate category is found where the combination of moderate to high wave energy and fine to medium sand results in a transitional beach type.

For beaches of a particular sediment size, wave conditions largely determine the beach state, which can change from reflective through intermediate to dissipative and back again with variations in weather. Reflective beaches can be cut back even by moderate wave action, whereas dissipative beaches are scoured only when swash levels are augmented by storms, a heavy swell, or exceptionally high tides. Particular events, such as a severe storm, can at least temporarily convert a dissipative beach into a reflective one, while prolonged fine weather can restore a dissipative beach profile: essentially the sequence of cut and fill described previously.

On the crenulate sandy beaches of the kind seen on the New South Wales coast all three types may co-exist within the beach compartment: reflective in the north, receiving relatively unrefracted ocean swell, intermediate in the centre, with nearshore bars dissected by rip current channels, and dissipative towards the southern end, where the waves have been much diminished by refraction over multiple sand bars, some of which are exposed at low tide. Within such a beach compartment it is possible for a moderate long swell to cut back reflective beaches towards one end at the same time as promoting accretion towards the other. There is thus no simple relationship between the three beach states and long-term trends of erosion or accretion; changes from one beach state to another occur during short-term cycles (e.g. cut and fill) operating on beaches that may show long-term progradation, stability or erosion.

This classification into reflective, dissipative and intermediate beach states was originally devised on microtidal coasts. There are complications where the tide range is larger, beach profiles exposed at low tide showing wider bars and troughs or low tide terraces. Masselink and Short (1993) examined features of beach morphology associated with the interaction of wave height and tide range, and introduced the concept of relative tide range (*RTR*), derived from actual tide range (*TR*) divided by breaker height (H_b):

$$RTR = TR/H_b$$

to distinguish categories of wave-dominated beach morphology as relative tide range increases. These range from reflective beaches ($\Omega < 2$, $RTR < 3$) to low tide terraces with rips ($\Omega < 2$, $RTR = 3$–7) then low tide terraces without rips ($\Omega < 2$, $RTR > 7$), from intermediate ($\Omega = 2$–5, $RTR < 7$) to low tide bar and rips ($\Omega = 2$–5, $RTR > 7$), and from barred dissipative ($\Omega > 5$, $RTR < 3$) to non-barred dissipative ($\Omega > 5$, $RTR > 7$) and ultra-dissipative ($\Omega > 2$, $RTR > 7$), with multiple lines of breakers moving in over a very wide low gradient profile, as on some Queensland beaches (e.g. Sarina Beach) and at Pendine Sands in Wales. These relationships are best seen on swash-dominated beaches, but there are further complications on drift-dominated beaches, where profiles vary laterally with such features as cusps and lobes, and bars formed at varying angles across the shore may be migrating laterally.

USE OF MODELS

Another approach to the study of beach dynamics is to set up scale models in tanks, in which waves and currents can be generated and the rise and fall of tides simulated, in order to test their effects on beach morphology and sedimentation. Such models can be used to demonstrate the relationships previously described: cut and fill sequences in response to changing wave conditions, the effects of wave refraction on beach outlines, and longshore drifting by waves arriving at an angle to the shore. The coastal and near-shore topography can be reduced to scale, as can waves and currents, but there are difficulties in scaling down natural sediment calibre to fit model conditions. Sand grains can be used in a model to represent a shingle beach, but representation of a sandy beach to scale may involve the use of silt, which has different physical properties and may not give the correct response of sand to wave and current action. Bearing in mind this limitation, scale models have been used to test hypotheses concerning, for example, the effects of introducing artificial structures, such as breakwaters and marinas, to the coast.

An alternative has been to develop mathematical models, which can provide computerised simulations of the response of coastal features to changing tide, wave and current conditions, using information obtained from surveys. Modelling is useful as a means of exploring process–response relationships, but coastal systems are complex and model predictions can prove unreliable. There is a trend towards field experiments, using temporary structures, as a prelude to coastal engineering works, and as will be seen in Chapter 5 there are advocates of the trial-and-error approach, particularly in beach nourishment projects.

EQUILIBRIUM BEACH PROFILES

The question of beaches attaining an equilibrium in plan has been mentioned (p. 45), and similar considerations apply to the evolution of equilibrium in beach profiles. Coastal geomorphologists and engineers have accepted the idea of an 'equilibrium beach' in terms of the attainment of a 'profile of equilibrium' (Bruun 1954), but there is some confusion over what this actually means in terms of the concept of equilibrium as defined below (Pilkey et al. 1993).

It is widely agreed that after a phase of sustained wave activity a beach profile develops a smooth, concave upward curve, the gradient of which depends partly on the grain size of the beach sediment, gravel beaches generally having steeper profiles than sandy beaches. This is evidently one kind of equilibrium beach profile, achieved by adjustment between the beach profile and the waves and currents at work on it.

Dean (1991) noted that such an equilibrium beach profile can be described as a concave upward curve:

$$h = Ax^{0.67}$$

where h is the still-water depth, x is the horizontal distance from the shoreline, and A is a dimensional shape parameter, the shape of which depends on the grain size of the beach material. Bodge (1992) preferred an exponential expression:

$$h = B\ (1 - e^{-kx})$$

in which B and k have dimensions of depth and inverse distance, which matches them more closely. These equations do approximate many surveyed beach profiles, especially those without nearshore sand bars, which are difficult to match with mathematical curves. It should be noted that many beach profiles are actually surveyed when calm conditions have become established following a phase of diminishing wave action.

The fact that these mathematical curves approximate surveyed profiles tells nothing about their stability, the morphogenic factors responsible, or their relationships with wave conditions. Indeed, as beach profiles steepen or flatten in response to changes in wave regime or granulometric composition they may show a variety of shapes similar to those generated by the formulae of Dean (1991) or Bodge (1992). It is possible that the smoothly curved profiles represent a condition that beaches attain when their granulometric composition has become adjusted to specific wave conditions. However, granulometric composition will be modified by gradual attrition as the result of wave agitation, and by sorting, and this will lead to profile changes unless mean grain size is maintained by a continual inflow of coarser sediment as the finer fraction is removed.

An equilibrium between the beach profile and nearshore processes may not have been attained before there is a change in processes, the sea becoming stormier or calmer, and a new adjustment begins.

The sequence of cut and fill observed on many beaches, with sediment removed during stormy phases and replaced during subsequent calmer weather, can be interpreted as a cyclic equilibrium to the extent that the profiles are restored by natural processes. This can occur over periods ranging from a few days or weeks (a storm and its aftermath) to a year or more (seasonal alternations).

Geomorphologists have used the term dynamic equilibrium to describe the condition where landforms or landform associations retain their shape after long periods, even though changes (uplift, erosion, deposition or subsidence) have occurred. On this basis, it could be said that a dynamic equilibrium exists on a beach when the profile and plan remain the same (perhaps with cyclic alternations) while the beach as a whole is either advancing (prograding) or receding (retrograding), but many would argue that this advance (by accretion) or retreat (by erosion) of the coastline really indicates disequilibrium. As will be shown in the next chapter, analyses of global patterns of beach change have indicated that over the past century most of the world's beaches have been retreating, some have been advancing, and a few have remained static or shown no net change. The supposition that a particular beach is (or has recently been) in some kind of equilibrium should therefore be treated with caution.

People concerned with beach management may find these equilibrium concepts somewhat academic, and certainly in need of refinement. When geomorphologists or engineers refer to beach equilibrium concepts they should indicate which kind of equilibrium they envisage, the time scale they are using, and whether they are dealing with the beach in plan or in profile, or the whole three-dimensional beach system.

In practical terms, beach management seeks to establish conditions that will ensure the maintenance of a beach. A cyclic equilibrium, whereby beach erosion during storms is compensated by fine-weather accretion, is satisfactory, especially if the phases of reduced beach coincide with phases of diminished beach use (i.e. the winter season) and the minimum beach width is then still sufficient to protect the backshore from erosion and damage to structures. In such a situation there are no net gains or losses of beach sediment over periods of more than a year or two. The suggestion that the attainment of a particular beach morphology, in plan or in profile, indicates the achievement of an equilibrium, with relatively stable relationships, is appealing, and attempts to do this will be discussed in Chapter 5.

BARS AND TROUGHS

Nearshore bars are typically built parallel, or almost parallel, to the shoreline. They may be intertidal, awash at low tide, or completely submerged. They are concentrations of sediment formed where the waves break, when material that is being carried shoreward meets that withdrawn from the beach by backwash. Such bars can be shaped in wave tank experiments,

Figure 19 Intertidal sand bars (ridge and runnel) on Mae Ramphong Beach, south-east Thailand

where it is found that their size and distance from the shore are related to the dimensions of the waves: higher waves build larger bars farther offshore. In calm weather, when constructive swash is more effective, the bars move closer to the shore, and become swash bars, flatter in profile, sometimes with a steeper shoreward advancing slope. Such bars may in due course become welded on to the beach face.

Nearshore bars cause waves to break, and at low tide a strong surf develops across them. They are often interrupted by transverse channels maintained by rip currents, which complete the nearshore water circulation by carrying the water driven on to the shore by breaking waves back seaward through the surf zone. Cuspate re-entrants form on sand and gravel beaches behind the heads of rip currents, where larger waves move in through water deepened by outflow scour (Komar 1983). Transverse, or finger, bars usually orientated at right-angles to the shoreline, are common off sandy beaches on low to moderate wave energy coasts, as in Florida and parts of North Queensland, and these are also produced by nearshore water circulations.

On sandy shores where the tide range is sufficient to expose a broad foreshore at low tide, systems of subdued ridges and troughs run parallel, or at a slight angle, to the coastline (Fig. 19). These are essentially flattened bars, formed where wave action is relatively weak, and are known as 'low-and-ball' or 'ridge-and-runnel' in Britain, where they are well developed in the intertidal zone on the Lancashire coast (Gresswell 1953). Their amplitude rarely exceeds a metre, and the ridge crests are often as much as 100 m apart. As the tide falls the troughs may be temporarily occupied by lagoons, which drain out by way of transverse channels (cut by rip currents). Once established, these features persist through many tidal cycles; they are associated with wave action as the tide rises and falls, and when the waves have been arriving at an angle to the shore the ridge pattern also runs obliquely. They are essentially the lower part of the beach system, and are used for recreation at low tide at seaside resorts such as Blackpool and Southport in Lancashire.

Occasionally the broad sandy intertidal zone is almost featureless, consisting of sand packed firmly by wave action. Such areas are used for sports such as sand yachting, as in The Netherlands and on the French Channel coast east of Dunkirk, and were even sites for motor racing some decades ago, as on Pendine Sands in South Wales.

FEATURES PRODUCED BY LONGSHORE DRIFTING

Reference has been made to longshore drifting as an influence on the shaping of beach outlines in plan (p. 43). Many beaches have been reorientated because sand or shingle eroded from one sector has drifted alongshore to accrete on another sector downdrift. This has led to the formation of spits and cuspate forelands, and progradation where drifting beach sediment has been intercepted by headlands and breakwaters.

Among many examples of spits that show progradation at their distal ends, matching erosion updrift, are those on Cape Cod in Massachusetts, USA, Blakeney Point in England, Pointe de la Coubre and Pointe d'Arcay off the French Atlantic coast, and Farewell Spit in the South Island of New Zealand. Barrier islands may have terminating spits: recession of the outer (west) coast of the German North Sea island of Sylt in recent decades has been accompanied by northward and southward drifting of sand to build spits at the northern and southern ends of the island. Sediment delivered to river mouths may be distributed alongshore as beaches and spits, as on the coastlines of the Rhône, Ebro and Mississippi deltas (Fig. 20).

Progradation of one side of a cuspate foreland as the result of the arrival of beach material eroded from the other side, leading to re-shaping and alongshore migration, has occurred on Capes Hatteras, Lookout, Fear and Kennedy, on the Atlantic coast of the United States; in each case erosion of the northern or eastern flank has been followed by accretion to the south (Shepard and Wanless 1971). The Darss Foreland on the German Baltic coast has prograded on its eastern flank as a result of erosion and transportation of sand from its western shore, and Zenkovich (1967) illustrated the similar evolution and migration of asymmetrical spits on the shores of

Figure 20 Recurved spit on the shores of the Mississippi delta, United States, formed by the westward drifting of sand delivered to the mouth of the river

the Sea of Azov. Other cuspate forelands showing such features include Dungeness, Winterton Ness and Benacre Ness in England (Bird and May 1976), the Magilligan Foreland in Northern Ireland (Carter 1988) and Cape Pitsunda on the Black Sea coast (Zenkovich 1973).

Many drift-dominated beaches (p. 44) include lobate protrusions which migrate alongshore. Lobes may be initiated by local convergence of beach drifting as the result of waves approaching obliquely, first from one direction, then the other, or they may be the result of local and temporary progradation beside a river mouth or lagoon outlet, moving on when the gap becomes sealed by deposition (Fig. 21). Some lobes form during occasional storms when waves arriving obliquely are strong enough to drive beach material round a headland to be deposited as a lobe on the lee side. However initiated, such lobes can move intermittently downdrift until they

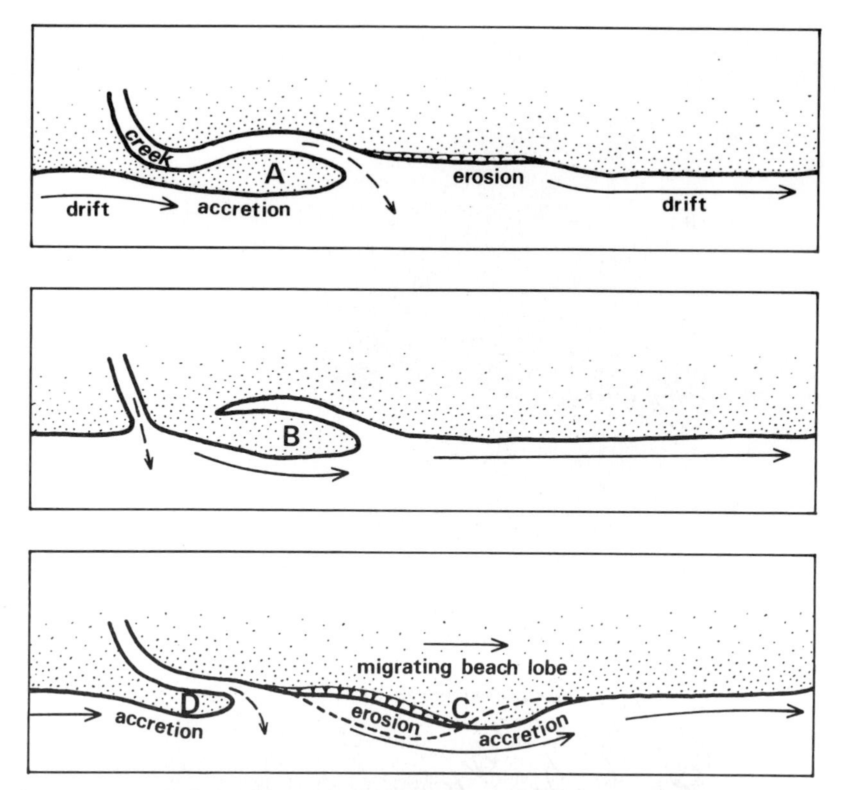

Figure 21 Evolution of a beach lobe (A) as the result of accretion updrift of a deflected creek, breaching to form a new outlet with the lobe downdrift (B), and subsequent migration of the lobe (C) alongshore. (D) is the next lobe

are intercepted alongside a headland or breakwater, or added to the end of a spit.

A succession of sandy lobes has formed and migrated along the beach system between Somers and Sandy Point in Westernport Bay, Australia, as the result of longshore drifting by waves from the south-west, as outlined and quantified in Fig. 15 (p. 42) (Bird 1985b). The passage of each lobe is marked by a phase of progradation, followed by erosion (Fig. 22) (p. 60).

On some coasts the beaches consist of a series of 'shingle trains' that move along the shore. In Southern England shingle is moving eastward along beaches in response to the prevailing south-westerly wave regime until it is interrupted by headlands and breakwaters, or until it accumulates in recurved spits, as at Hurst Castle in Hampshire, or cuspate forelands, as at Dungeness.

On coasts exposed to ocean swell longshore drifting is produced only if the swell is incompletely refracted, so that it arrives at an angle to the

Figure 22 Granite boulders dumped in an attempt to halt erosion at Somers Yacht Club, Westernport Bay, Australia (location of Fig. 12), which had been built on a lobe (Fig. 21) that has moved on along the coast. The dotted line indicates the shoreline in 1970

beach, or if waves generated by onshore winds move at an angle through the swell to reach the shoreline obliquely. On the Adelaide coast in South Australia, for example, south-westerly waves move sand northward to accumulate on the prograding beach at Largs Bay, where longshore drifting is intercepted on the southern side of the Outer Harbour breakwater (Fig. 74, p. 142).

Where longshore drifting of sand and gravel is interrupted by a protruding headland there is accretion on the updrift side, as at Point Dume in California, Apam in Ghana and Cape Wom on the north coast of New Guinea. Such a beach will continue to prograde until sand or gravel can by-pass the promontory to reach the downdrift shore.

This effect has been reproduced at many places where jetties or breakwaters* have been constructed, usually to stabilise a river mouth or

* The terms jetty, breakwater and pier can be defined as follows. A jetty is a solid structure built more or less at right angles to the coastline or on either side of a river mouth or lagoon entrance, a breakwater is an often curved structure designed to impede wave action so as to shelter a harbour or protect a sector of coastline, and a pier is an open structure on multiple supports, usually designed to permit ships to berth, and beneath it waves, currents and drifting sediment pass almost unimpeded. However, the terms are widely interchanged or regarded as synonymous.

lagoon outlet, or shelter a harbour from winds and waves. There are many examples of longshore drifting of beach sediment being interrupted by breakwaters, resulting in progradation updrift and erosion downdrift of the intercepting structure. This has happened, for example, at Santa Barbara and Redondo in California, South Lake Worth in Florida, Sochi on the Black Sea coast, Lagos in Nigeria, Durban in South Africa and Madras in India. In south-east England, longshore drifting is generally southward on the East Anglian coast, where beaches have prograded on the northern side of harbour breakwaters at Yarmouth, Lowestoft and Southwold, while on the Channel coast, eastward longshore drifting has led to accumulations on the western side of breakwaters such as The Cobb at Lyme Regis, Black Rock marina at Brighton, and at Newhaven, Rye and Folkestone. At Studland, in Dorset, the beach has prograded south of a training wall built in 1924 to prevent the northward drift of nearshore sand from shoaling the entrance to Poole Harbour (Bird and May 1976). Among many other examples are the interception of southward-drifting sand by the breakwaters at Praia da Barra near the Aveiro Lagoon and at Figuera da Foz, both in Portugal, and eastward drifting sand alongside the jetty built at the mouth of the Vridi Canal on the Ivory Coast when it was cut in 1950. In Australia, northward-drifting sand has accumulated along the southern sides of breakwaters at Port Adelaide, and at Tweed Heads in northern New South Wales.

In some places progradation has occurred on both sides of protruding breakwaters, indicating either an alternation of longshore drifting or a predominance of shoreward drifting on to beaches on either side of the outflow. There has been sand accretion on both sides of the breakwaters built at Lakes Entrance in south-east Australia to stabilise the artificial outlet from the Gippsland Lakes cut in 1889 (Fig. 23). Similar accretion has taken place on both sides of paired breakwaters at Onslow in north-western Australia, Rogue River and Siuslaw River on the Oregon coast, Newport in California, Ijmuiden on the Dutch coast, and the Swina Inlet in Poland.

There are also examples of beach accretion being induced in the lee of a breakwater, as at Warrnambool in Victoria, Australia, where shoreward drifting of fine sand occurred from the sea floor into the area that became more sheltered after the breakwater was constructed (Bird 1993a).

Accretion has taken place alongside some tidal inlets (lagoon entrances) where no breakwaters have been built, the transverse ebb and flow of currents having had the same effect as an intercepting breakwater. There is usually some deflection of an unprotected outflow channel by longshore drifting, and if this continues and the gap becomes sealed off by deposition, the accreted foreland may become a lobe, which then moves on along the shore (see Fig. 21).

On the north-east coast of Port Phillip Bay, Australia, where seasonal patterns of longshore drifting occur (p. 40), breakwaters built to form boat

Figure 23 Sand accretion on either side of the breakwaters built beside an artificial entrance to a coastal lagoon system (the Gippsland Lakes) at Lakes Entrance, Victoria, Australia

harbours have intercepted the southward drift in winter in a situation whence it cannot be returned northward in summer. The harbours thus become filled with sand and the adjacent beaches are depleted. An example of this is seen at Sandringham, near Melbourne, where the beach has prograded within a harbour sheltered by breakwater construction, which unexpectedly formed a sand trap (Fig. 24). This kind of problem indicates the difficulties that arise when coastal constructions interfere with longshore drifting of beach sediment. Knowledge of sediment sources and patterns of longshore drifting is essential if constructions are to be designed in such a way as to conserve the natural processes of erosion and deposition on adjacent shores.

TRACING BEACH SEDIMENT FLOW

Patterns of longshore drifting on beaches can be deduced from such indications as accretion alongside headlands or breakwaters, the migration of beach lobes, or the deflection of river mouths and lagoon outlets (Schwartz et al. 1985). Pebbles of an unusual rock type or specific mineral sands may act as natural tracers (see below) indicating longshore drifting from a source area such as a cliff outcrop or river mouth: as has been noted (p. 18) the mineral augite appears in the Ninety Mile Beach, in south-eastern Australia, at the mouth of the Snowy River and extends eastward, the movement of this fluvially supplied mineral indicating longshore drifting in that direction.

Figure 24 Sand began to accumulate in Sandringham Harbour, Port Phillip Bay, Australia, after the construction of the sheltering breakwater. See Fig. 46. X marks the location of sand extraction in 1975 to nourish Hampton Beach (p. 100)

Usually at least some of the lateral movement of sand and gravel takes place along the nearshore zone, or by way of migrating exchanges between the beach and the nearshore zone, with subsequent delivery to the beach downdrift. Some of the indications of longshore drifting mentioned above may be misleading where patterns of beach accretion result partly from sediment movement in from the sea floor, rather than alongshore.

A more accurate picture of patterns of beach sediment movement can be obtained by the use of introduced tracers. These are deposited on a beach or in the nearshore zone at a particular point or along a selected profile, and surveys are made subsequently to see where they have gone. If it is to reproduce the movements of the natural beach sediment, a tracer must consist of particles similar in size and shape, and with similar hardness and the same specific gravity as the sediment already present; it must also be identifiable after it has moved along a beach or across the sea floor. Some or all of the tracer may become buried within a beach, and a proportion may be lost seaward from the nearshore zone.

Some tracer projects have used sand containing minerals that were not naturally present on the beach. These were injected at a particular point, and surveys then made to discover where they went. The mineralogical

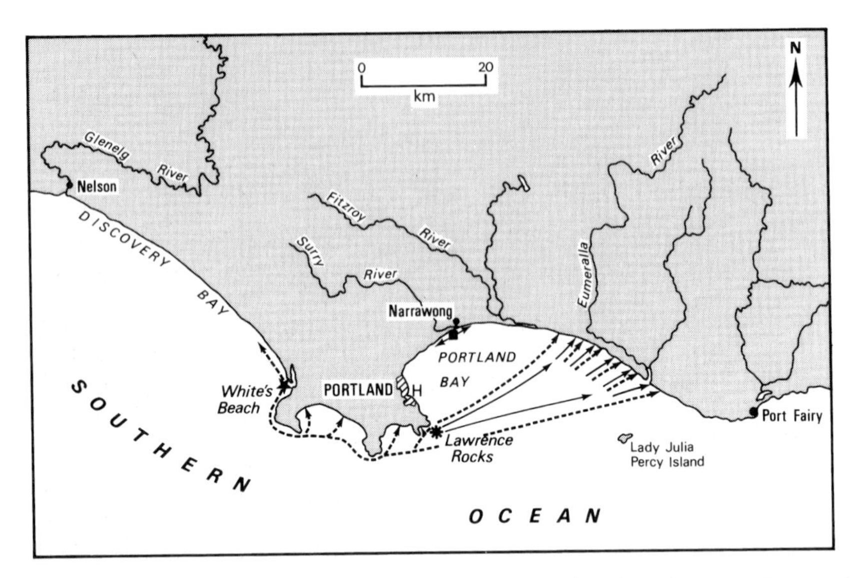

Figure 25 Pathways of sand movement on to beaches in Discovery Bay and Portland Bay, Australia, after mineral tracer was injected at Whites Beach, Lawrence Rocks and Narrawong by Baker (1956). The aim was to determine whether there would be significant sand movement into the harbour that was subsequently built at Portland (H). Minor accretion has occurred, but it is of fine sand washed up from the nearby sea floor rather than coarse beach sand moving along the coast

method depends on a complete preliminary inventory of the minerals present in the beach. Baker (1956) had some success using alien 'heavy minerals', such as iron pyrites, which he introduced to beaches of quartz and calcarenite sand on the west coast of Victoria, Australia, as a prelude to surveys to discover the patterns of sand movement alongshore and across Portland Bay, where a harbour was to be constructed (Fig. 25). The project indicated correctly that most of the sand moving along the coast was passing out across the sea floor, and would not accumulate in the harbour, but there are, nevertheless, risks in using tracers of higher specific gravity than the natural beach sediment, because the introduced minerals may not reproduce the behaviour of the natural beach sediment. Lighter grains are moved more frequently and further than heavy grains, and their net long-term drifting may not be in the same direction.

Sand or gravel that is naturally or artificially coloured can be used as tracer, but there are difficulties in observing coloured grains when they form only a small proportion of the sediment on a beach. More effective is the use of natural or artificial sand or gravel coated with a colloidal substance containing a fluorescent dye (Ingle 1966). Sand or gravel labelled in this way is dumped on a beach (Fig. 26), and its subsequent movement

Figure 26 Laying fluorescent material (dotted line) across a shingle beach for a tracing experiment at Blakeney Point, England

traced by locating the dyed material at night, using an ultra-violet lamp, or sand samples can be examined under ultra-violet light in a darkroom, when the fluorescent stain stands out brightly.

Another method of tracing sediment flow depends on the use of radioactive materials, which can be followed by means of a detection meter (a geiger counter). Artificial sand can be made from soda glass containing scandium oxide, which is ground down to the appropriate size and shape for use as a tracer. It is then placed in a nuclear reactor to acquire the radioactive isotope scandium 46, which has a half-life (i.e. time taken for the radioactivity to diminish to half its original strength) of 85 days, which means that it can still be detected three or four months after it has been introduced. Gravels can be traced by using granules, pebbles or cobbles that have been taken from the beach and labelled with a radioactive substance, or artificial materials such as concrete in which radioactive material has been embedded. After the tracer has been deposited, surveys are made by carrying a detection meter over a beach, or dragging it across the sea floor mounted on a sledge. The location and intensity of radioactivity is thus mapped and paths of sediment flow can be deduced.

Radioactive tracers are expensive and a health hazard, but they provide a good means of tracing sediment movement, for the tracer can be detected even when it has been buried in a beach, where coloured or fluorescent tracers would not be visible. However, fluorescent tracers are cheaper and safer, and often more durable for long-term projects, whereas a radioactive tracer permits only one project, and cannot be used again until all the tracer used in the first project has disappeared. Different colours of paint or fluorescent dye can be used to trace sediment movement from several places at the same time.

Tracers have been used to estimate rates and quantities of beach sediment movement. On Sandy Hook, New Jersey, Yasso (1965) introduced four grain size classes of fluorescent sand, and subsequent sampling on a profile 100 feet downdrift showed that the smallest particles (0.59–0.70 mm) arrived first, and the next class (0.70–0.84 mm) soon afterwards, the maximum rates of flow being 61 and 79 cm/min, respectively.

If the predominant direction of longshore drifting is known it is possible to estimate the volume of sediment moving along the beach by introducing a standard quantity of tracer at regular intervals to a particular 'injection point', and taking samples at another point downdrift to measure the concentration of tracer passing by. As the concentration of tracer downdrift is proportional to the rate at which the sediment is moving, the quantities in transit can be measured. Jolliffe (1961) applied this tracer concentration method successfully to the measurement of rates of longshore drifting on shingle beaches on the south coast of England, using pebbles coated with a fluorescent substance. It is more difficult to trace sand movement by this method because of the vast number of grains involved, but sand movement on the Caucasian Black Sea coast has been measured using fluorescent tracer sand at a dilution of one grain in 10 million.

LATERAL VARIATIONS IN BEACH MATERIAL

Some beaches show a gradation from fine to coarse sediment in one or other direction along the shore. The grain size composition of beach material may vary laterally, particularly in the vicinity of eroding headlands, where the proportion of locally derived coarse material may be high, and near river mouths, where a larger proportion of coarse fluvial sediment is likely to be present. Lateral grading by size has been reported from many beaches, but there are also beaches showing alongshore grading in terms of the shape or specific gravity of sediments.

There is no single, simple explanation for lateral grading of beaches by particle size, and several hypotheses have been put forward. One is that there has been longshore sorting of beach material by breaking waves and nearshore currents, so that a beach which initially had particles of various sizes shows selection in the course of longshore drifting, the finer grains being moved further because they are more easily mobilised by waves and transported by associated currents. Alternatively, there may be selective removal of the finer grains from the beach, either because they are blown

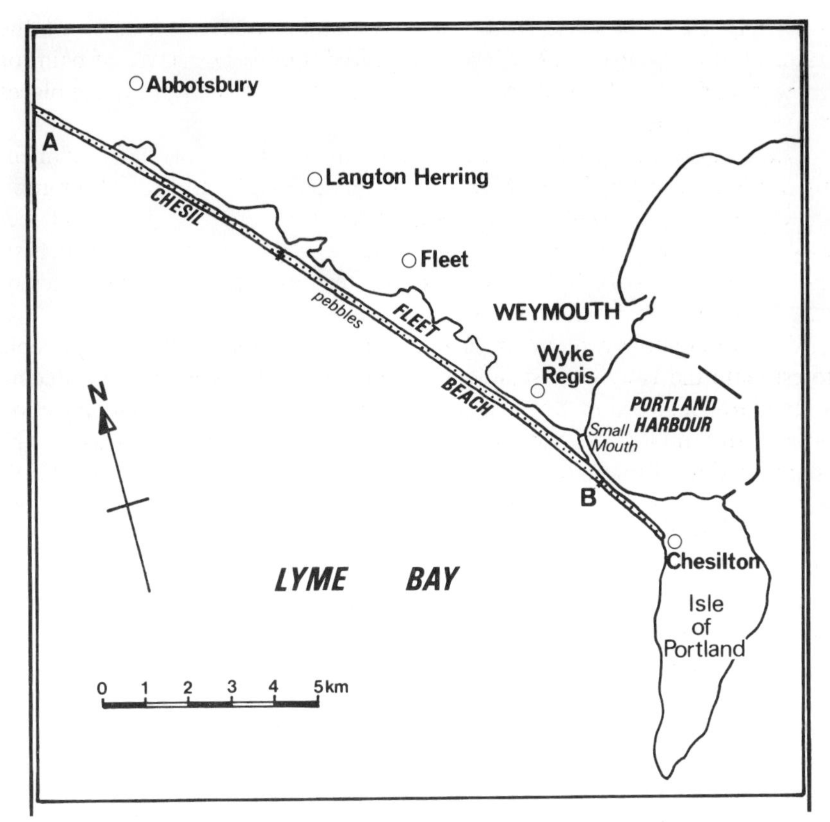

Figure 27 Chesil Beach, on the south coast of England. A and B indicate the locations of the photographs shown in Fig. 28

offshore or onshore by wind action, or because of withdrawal seaward by backwash, the extent of such removal diminishing alongshore (Komar 1976).

Chesil Beach, on the south coast of England (Fig. 27), consists mainly of flint pebbles, and grades from granules on the beach at the western end, near Bridport, through to pebbles and cobbles towards the south-eastern end, near Portland Bill (Fig. 28). This gradation is correlated with a lateral increase in wave energy, with larger waves coming in through deeper water to build the higher and coarser beach at the more exposed south-eastern end. It is not clear how this contrast in wave energy actually achieved lateral sorting, but longshore drifting may have contributed by edging the finer particles from higher to lower wave energy sectors. Chesil Beach is still receiving small amounts of shingle from cliff erosion to the west and to the south-east, where adjacent cliffs of Portland limestone yield cobbles and pebbles (mainly from quarries), which become mixed with the flint

Figure 28 Lateral grading on Chesil Beach. The scale in both photographs is a 1 foot (approximately 30 cm) ruler. On the left is the beach near the western end (A in Fig. 27), where the mean pebble diameter is between 1 and 2 cm, and on the right the beach towards the south-eastern end (B in Fig. 27), where the mean pebble diameter is about 5 cm

shingle (Carr and Blackley 1974), and it seems that pebbles of particular dimensions are then moved quickly to the appropriate size sectors by longshore drifting, and there retained.

Bascom (1951) related a similar gradation in beach sand particle size on the shores of Half Moon Bay, California (p. 40), to variation in degree of exposure to refracted ocean swell coming in from the north-west, coarse sand on the exposed sector grading to fine sand in the lee of a headland at the northern end. Again there is an implication that lateral sorting developed as sand was edged from a higher to a lower wave energy sector.

Another possible explanation of lateral grading on beaches is where longshore drifting is stronger in one direction than the other, as the result of contrasts in the height or frequency of incident waves. The larger waves carry the whole range of available coarse to fine particles in one direction, but the smaller waves take back only the finer material (Fig. 29). Chesil Beach could have become laterally sorted in this way (Jolliffe 1964).

Lateral gradation of particle size may alternatively be the outcome of wearing and attrition of sediment derived from a particular sector of the coast as longshore drifting carries it away from the source. This may explain the graded beach on the coast of Hawke Bay, New Zealand, which is supplied with gravelly material by the Mohaka River, and shows a

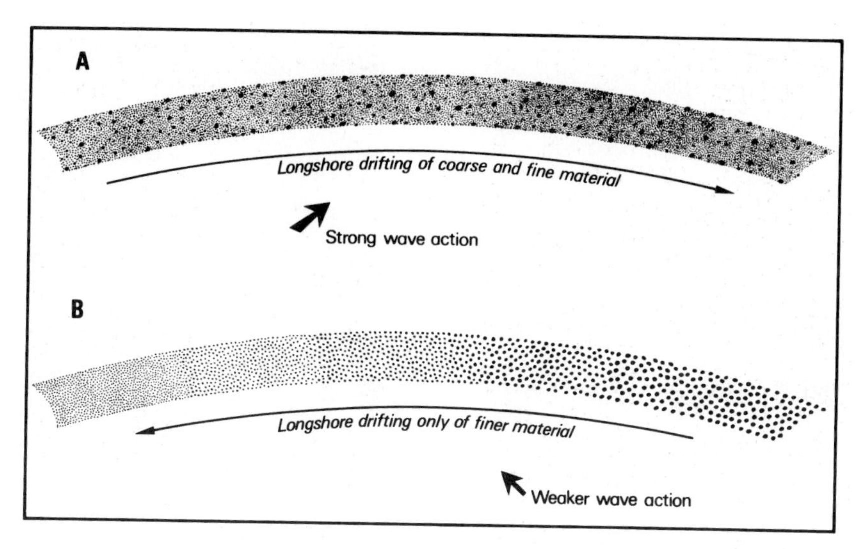

Figure 29 Evolution of lateral grading of a beach by alternations of strong wave action from one direction and weaker wave action from the other

gradual reduction of mean particle size from pebbles to granules, then coarse to fine sand, along the shore eastward from the mouth of the river (Bird 1996).

Lateral variation may also occur in the mineral composition of beach material. This is also likely to be the outcome of sorting, the heavier minerals being concentrated selectively by longshore drifting, but it may alternatively reflect the pattern of sources. In Encounter Bay on the South Australian coast the south-eastern end of the beach, near Kingston, consists almost entirely of calcareous sand (almost 90% calcium carbonate) washed up from the sea floor, but the proportion of quartz sand increases north-westwards to Goolwa, at the mouth of the Murray (calcium carbonate content less than 10%). This lateral variation is the outcome of earlier deposition of fluvial quartzose material on the sea floor off the mouth of the Murray during Pleistocene low sea level phases, so that the sand swept onshore is there dominated by quartz rather than the calcareous sand of marine origin further along the bay shore.

PROGRADING BEACHES

In Chapter 1 it was noted that if beaches receive more sediment from various sources than they lose onshore, offshore or alongshore they become higher and wider, and prograde seaward. The dynamics of a beach system are determined by the balance of nourishment from fluvial sources, cliff and

Figure 30 Prograded beach at Malindi, Kenya. The vegetation advanced on to the backshore as the beach was widened by sand accretion

rocky foreshore yields, the sea floor, or dunes blown from the hinterland against alongshore and offshore losses, the removal of sand by wind to build landward dunes, and the washing of sand into estuaries and tidal inlets (see Fig. 2).

The alternation of cut and fill on a shore receiving abundant sediment results in progradation, usually accompanied by the formation of a succession of parallel beach ridges or foredunes (p. 75). Backshore vegetation spreading forward on to the beach indicates that progradation has been taking place (Fig. 30), especially where the vegetation canopy declines smoothly seaward from trees through shrubs to grassy communities, a zonation that indicates that a plant succession is accompanying the deposition of new terrain.

Beach ridges can also be formed by the successive addition of spits that have grown parallel to the coast. Historical maps of South Haven Peninsula on the Dorset coast indicate an evolution of this kind (Robinson 1955), and the peninsula at Falsterbo, in south-west Sweden, also prograded in this way.

Beach progradation on the Danish island of Kyholm resulted from the disappearance of seagrass meadows in the surrounding waters and the ensuing shoreward movement of fine sand that had previously been retained on the sea floor by this vegetation (Christiansen et al. 1981). Beach progradation may be induced by human activities, notably where artificial

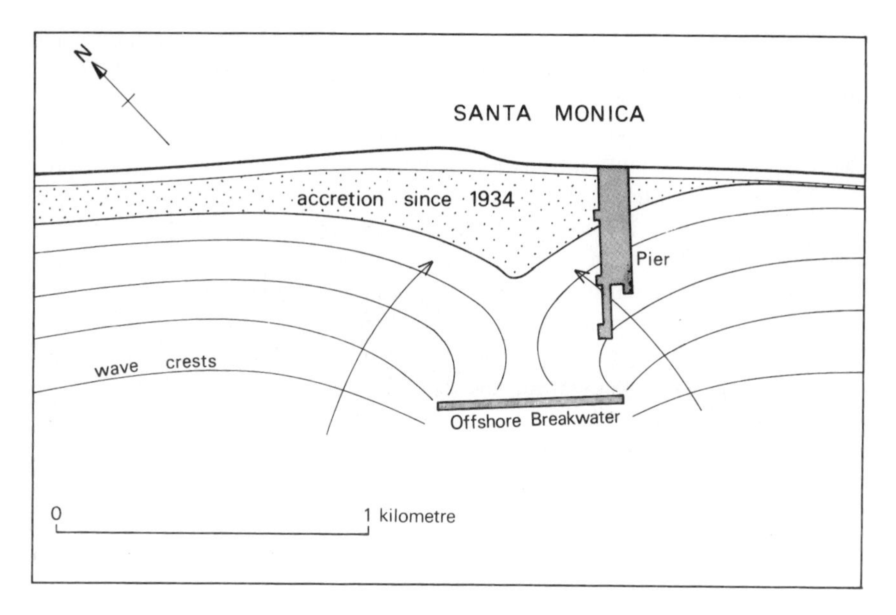

Figure 31 Formation of a cuspate beach in the lee of an offshore breakwater at Santa Monica, California, as the result of convergent drifting with waves refracted by the breakwater

structures have been built on or near the coast, or where sediment yields have increased as the result of hinterland deforestation, mining or other impacts. Reference has been made to progradation alongside breakwaters that have intercepted the longshore drifting of beach material, and local progradation has also taken place in the lee of offshore structures. An example is the sandy foreland that developed behind the offshore breakwater built in 1934 at Santa Monica, California (Fig. 31). Similar local progradation has occurred behind the offshore breakwaters built to protect Italian beaches, notably between Rimini and Venice on the Adriatic coastline. Progradation of a beach can also be induced in the lee of a wrecked ship close to the shore, as at Sukhumi on the Caucasian Black Sea coast (Zenkovich 1967).

ERODING BEACHES

Beaches are eroded when they lose more sediment alongshore, offshore or to the hinterland than they receive from the various sources mentioned in Chapter 1. Processes that lead to beach erosion, mentioned in this chapter, include destructive wave action in stormy periods, the effects of storm surges, and the depletion of beach sediments by weathering and winnowing.

The idea that beaches should naturally be stable, having attained some kind of equilibrium with the processes at work on them, is difficult to sustain from geological evidence. The geological record indicates that land areas have been planed off by the sea during prolonged periods of erosion, then buried by accumulations of sediment from various sources, particularly rivers, and by deposits of organic origin, such as limestones. Beaches form only a very minor proportion of the world's sedimentary formations; they have been preserved, for example, as parts of prograded barrier formations within Tertiary deposits in Texas, but they are not prominent in most depositional systems (Davis 1983). This is because they have been transient features, formed and reworked by wave energy, on coasts that are destined either to retreat as marine planation proceeds, reducing land masses ultimately to surfaces planed down to wave base, the limit of wave erosion, or to be submerged and dispersed by the rising sea, and buried by younger sedimentation. It could be argued that beach erosion is natural, and that the attempt to preserve existing beaches will become increasingly futile.

The causes of beach erosion will be examined in more detail in the following chapter.

Chapter Three

Causes of beach erosion

MEASURING BEACH CHANGES

Changes in the plan and profile of beaches can be measured by making repeated surveys along and across them, using conventional methods with instruments such as a level or theodolite. Traditionally, beach profiles have been surveyed at right angles to the coastline, from backshore datum points (marked so that they can be found for subsequent surveys) down to the low tide line, and some distance seaward in shallow water. These can be used to monitor the advance or retreat of the coastline, supplementing information from series of dated air photographs. When they are linked by alongshore surveys, gains and losses of land area on the coast and in the intertidal zone can also be measured. Beach budgets can then be computed by multiplying the mean cross-sectional area of neighbouring profiles (using arbitrarily defined basal and rear planes) by the intervening longshore distance to obtain volumetric changes (p. 42). Alternatively, it is possible to make successive contour maps of the beach surface, and to determine patterns and volumes of gain or loss between each survey from these.

Various tracking meters and vehicles have been developed to accelerate beach surveying, and an important recent innovation has been the use of the satellite-based GPS (Global Positioning System), particularly on beaches firm enough for vehicle-based surveys, as on the Gulf Coast in Texas (Morton et al. 1993). On the Adelaide coast, in South Australia, GIS (Geographical Information Systems) have been used to generate serial contour maps showing the pattern of beach surface gains and losses as a basis for planning further beach nourishment programmes (Fotheringham and Goodwins 1990).

SHORT-TERM EROSION AND ACCRETION

It has been shown in Chapter 2 that beaches advance or retreat with alternations of relatively gentle constructive wave action and episodes of stormy erosive waves, with variations in rates of sediment supply from rivers and cliffs, in the vigour of winds that supply or remove sandy sediment, and in the direction, strength and duration of longshore drifting.

There are both short-term changes, over periods of up to a year or a few years, and longer-term changes measurable over decades and centuries. The former include changes that correspond with tidal phenomena. For example the sequence of increasingly high tides from neaps to springs simulates a marine transgression, and is often an erosional phase; it is followed by a sequence of diminishing high tides from springs to neaps, effectively a regression of the sea, which is often accompanied by secular beach accretion. Seasonal changes occur where there is a strong contrast between a wet and a dry season, between stronger winter and gentler summer wind and wave action, or between the direction of approach of dominant winds in different seasons.

Short-term changes may be cyclic, the beach plan and profile regaining their previous shapes after intervals of modification by erosion or accretion, or there may be a definite trend towards erosion or progradation, evident when changes over longer periods are considered. Indications of beach erosion include cliffed backshore dunes (whereas prograding beaches are backed by beach ridges and incipient foredunes), truncated vegetation zonations (whereas prograding beaches are backed by tree canopies descending to beach level, or by shrub and grass zones on recently formed sandy terrain), or exposures of beach rock (p. 39). Measurements of changes over a longer period, based on air photographs and satellite imagery, use time scales of a few years or decades, while historical maps and charts can be used to assess changes over a century or more. On long-settled coasts changes can be determined from historical and archaeological evidence, as around the Mediterranean Sea, where it is locally possible to document the advance or retreat of the coastline over at least two thousand years. Changes since the Late Quaternary marine transgression which brought the sea to approximately its present level (about 6000 years ago on stable coastlines) can be deduced from stratigraphical evidence of coastal deposits, with dating of sediments that indicate past shoreline positions (e.g. Hough and Menard 1956).

Between 1976 and 1984 the Commission on the Coastal Environment (International Geographical Union) assembled world-wide evidence of coastline changes over the preceding century, and found that beach erosion has become widespread. More than 70% by length of beach-fringed coastline had retreated over this period, less than 10% having advanced (prograded), and the balance having either remained stable or shown alternations with no net gain or loss (Bird 1985a).

Beach erosion has been reported from many coasts. Seaside resorts such as Deauville in France, Miami in Florida ánd Surfers Paradise in Australia are among the many that have suffered when storms have removed the

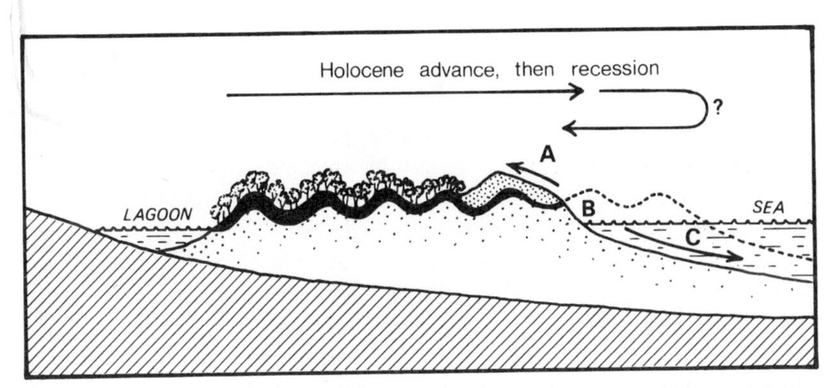

Figure 32 Diagram showing evolution of barrier coastlines where Holocene progradation formed a succession of parallel ridges, but has been followed by recession, with losses of sand landward into spilling dunes (A), alongshore (B) and offshore (C)

beaches that attracted their visitors. In such cases recurrent 'cut' has exceeded subsequent 'fill', only part of the sand removed by storm wave scour being returned to the beach by natural processes in calmer weather. There has been retreat of sandy coastlines in many parts of the world during the past several decades, and in many places for a much longer period. Recession is in progress even on the shores of barriers that were previously prograded by deposition (Fig. 32), for example along the Ninety Mile Beach in south-eastern Australia, where progradation during the past century has been confined to sectors on either side of the Lakes Entrance harbour jetties (see Fig. 23). It is quite difficult to find sectors of naturally prograding sandy beach, whereas receding sandy shorelines are extensive. Only a few seaside resorts have had their beaches substantially widened by natural accretion: examples are Seaside in Oregon and Malindi in Kenya.

Beach management must therefore bear in mind the modern prevalence of beach erosion: stable or prograding beaches are unusual.

THE MODERN PREVALENCE OF BEACH EROSION

Beach erosion occurs where the various losses of beach material exceed the gains (see Fig. 2). As the volume of beach material diminishes, the beach face is lowered and cut back. Beaches that line cliffed or rocky coasts become narrower, thinner and discontinuous as losses proceed, and may eventually disappear, while on depositional coasts the backshore becomes cliffed as dunes, beach ridges or other previously deposited landforms are cut back (Fig. 33). The rate of retreat of the high tide line, which is often also the seaward boundary of terrestrial vegetation communities, can be measured by comparing dated sequences of maps and charts, or air and

Figure 33 Sectors of the Ninety Mile Beach are backed by a receding cliff cut into backshore dunes, behind a beach lowered by erosion (Fig. 32)

ground photographs. Mean annual recession rates have been generally small (less than 1 m per year), but some beaches have retreated by up to 40 m per year.

Investigating the causes of beach erosion, the Commission on the Coastal Environment found that there was no single, simple and universal explanation (Bird 1985a). Some 20 factors (Table 3) have been identified as having initiated or accelerated beach erosion, their relative importance varying from one coastal sector to another. Analyses of particular eroding beaches in terms of these 20 factors have shown that usually more than one has contributed to the onset or intensification of erosion. Management of an eroding beach poses many problems, and it is necessary to understand the causes of erosion before seeking remedies. Each of the possible causal factors should be considered, and those found to be applicable ranked in importance as a means of comprehending the eroding beach system and considering what action should be taken.

CAUSES OF BEACH EROSION

Submergence and increased wave attack

Coastal submergence may be due to a rise of sea level, coastal or nearshore land subsidence, or some combination of land and sea movement that results

Table 3 Causes of beach erosion

1. Submergence and increased wave attack as the result of a rise in sea level or coastal or nearshore land subsidence: the 'Bruun Rule'.
2. Diminution of fluvial sand and shingle supply to the coast as a result of reduced runoff or sediment yield from a river catchment (e.g. because of a lower rainfall, or dam construction leading to sand entrapment in reservoirs, or successful revegetation and soil conservation works), or the natural or artificial diversion of a river outlet, or because of the removal of much or all of the weathered mantle from slopes in the river catchments, exposing bare rock that yields little or no sediment to runoff.
3. Reduction in sand and shingle supply eroded from cliffs or shore outcrops (e.g. because of diminished runoff, stabilisation of landslides, a decline in the strength and frequency of wave attack, or the building of sea walls to halt cliff recession), or the exposure by cliff recession of formations (such as massive resistant rock, or soft silts and clays) that do not yield beach-forming sediment.
4. Reduction of sand supply to the shore where dunes that had been moving from inland are stabilised, either by natural vegetation colonisation or by conservation works, or where the sand supply from this source has run out.
5. Diminution of sand and shingle supply washed in by waves and currents from the adjacent sea floor, either because the supply has declined (e.g. where ecological changes have reduced the production of shelly or other biogenic material), because the transverse profile has attained a concave or steeply declining form which no longer permits shoreward drifting, or because such drifting has been impeded by increased growth of seagrasses or other marine vegetation.
6. Removal of sand and shingle from the beach by quarrying or the extraction of mineral deposits, or losses from intensively used recreational beaches, for example as the result of beach cleaning operations.
7. Increased wave energy reaching the shore because of deepening of nearshore water (e.g. where a bar or shoal has drifted away, where seagrass vegetation has disappeared, where dredging has taken place, or where the sea floor has subsided).
8. Reduction in sand and shingle supply from alongshore sources as the result of interception (e.g. by a constructed breakwater or groynes) or interruption by the growth of a fringing coral reef or some other depositional feature.
9. Increased losses of sand and shingle alongshore as a result of a change in the angle of incidence of waves (e.g. as the result of the growth or removal of a shoal, reef, island or foreland, or because of breakwater construction or a change in wind regime).
10. Intensification of obliquely incident wave attack as the result of the lowering of the beach face on an adjacent sector (e.g. as the result of dredging, or scour due to wave reflection induced by sea wall or boulder rampart construction or beach mining).
11. Increased losses of sand from the beach to the backshore and hinterland areas by onshore wind action, notably where backshore dunes have lost their retaining vegetation cover and drifted inland, lowering the terrain immediately behind the beach and thus reducing the volume of sand to be removed to achieve coastline recession.

continued overleaf

Table 3 (*continued*)

12. Increased wave attack due to a climatic change that has produced a higher frequency, duration, or severity of storms in coastal waters.
13. Diminution in the calibre of beach and nearshore material as the result of attrition by wave agitation, leading to winnowing and losses of increasingly fine sediment from the beach, either landward into backshore dunes or seaward to bars and bottom deposits.
14. Diminution of beach volume by weathering, solution, attrition or impaction (e.g. by heavy vehicle traffic), resulting in lowering of the beach face and coastline recession.
15. Increased scour by waves reflected from backshore structures, such as sea walls or boulder ramparts, leading to reduction of the beach that fronted them.
16. Migration of beach lobes or forelands as the result of longshore drifting – there is progradation as these features arrive at a point on the beach, followed by erosion as they move away downdrift.
17. A rise in the water table within the beach, due to increased rainfall or local drainage modification, rendering the beach sand wetter and thus more readily eroded.
18. Increased losses of beach material due to outwashing by streams or from drains carrying an augmented discharge of water, either because of increased rainfall or from melting snow or ice, or because of natural or artificial modifications (such as urbanisation) in the catchments that have increased runoff through beaches.
19. Intensification of wave action where tide range has diminished, as on the shores of a bay or lagoon that has been partly enclosed by natural or artificial structures, impeding tidal ventilation.
20. On arctic coast beaches the removal of a protective sea ice fringe by melting, so that waves reach the beach (e.g. for a longer summer period).

in the sea standing higher relative to the land. It is likely to lead to beach erosion, for a sea level rise usually results in the deepening of nearshore waters, allowing larger waves to reach the shore and erode the beach face, thus withdrawing sand or gravel to the sea floor. Bruun (1962) suggested that on beaches which had attained a 'profile of equilibrium' (which he took as a condition where they were neither gaining nor losing sediment over a specified period), a sea level rise would cause erosion of the upper beach, and transference of sand or gravel from the beach to the adjacent sea floor would in due course restore the previous transverse profile in relation to the higher sea level (Fig. 34). The transverse profile would thus migrate upward and landward, the coastline retreating further than it would have done with an equivalent sea level rise on a rocky shore of similar profile.

This has become known as the Bruun Rule. It has been reproduced in wave tank experiments (Schwartz 1967), and is supported by observations of the backshore erosion and nearshore deposition that have taken place on

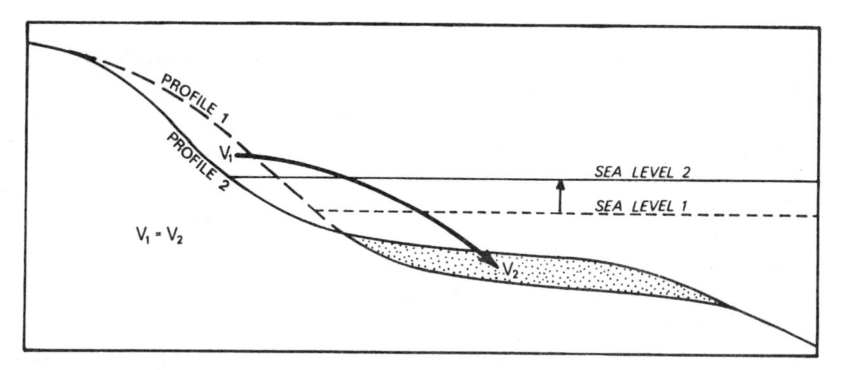

Figure 34 The Bruun Rule states that a phase of sea level rise on a sedimentary coast will be accompanied by recession of the coastline as a volume of sediment is transferred from the backshore to the nearshore zone in such a way as to restore the transverse profile. For discussion see pp. 80–82.

beaches around the Great Lakes during and after each episode of rising water level (Fig. 35). As a rule of thumb it is predicted that if there are no gains or losses from the hinterland or alongshore, beach erosion will cause the coastline to retreat by 50 to 100 times the dimensions of the rise in sea level, i.e. a 1 m rise will result in the high tide shoreline being cut back 50–100 m.

Beach erosion has certainly become widespread on coasts where the sea has been rising because land subsidence is in progress, as on the Gulf and Atlantic coasts of the United States, notably in Louisiana west of the Mississippi delta. Land subsidence has contributed to beach erosion on the northern Adriatic coast of Italy, as on the Lidi of Venice, and beaches were cut back suddenly on sectors of the Alaskan coastline that subsided during the 1964 earthquake.

Evidence from tide gauge records suggests that there has been a world-wide sea level rise of about a millimetre per year during the past few decades, offset on some coasts by equal or greater land uplift, and probably varying also in relation to geophysical factors that complicate the surface topography of the oceans (Pirazzoli 1991). The poor geographical distribution of reliable tide gauge records has impeded firm conclusions concerning global trends of sea level, but this situation is improving with the setting up of the GLOSS (Global Sea Level Observing System) network of about 300 well-distributed tide gauges (Woodworth 1991). If there has indeed been a sea level rise, the Bruun Rule may provide at least a partial explanation for the modern prevalence of beach erosion.

Although the Bruun Rule appears valid in general terms, there has been much discussion of problems which complicate its applicability to precise forecasting of the extent of coastline retreat after a sea level rise (e.g. SCOR Working Group 1991). The following eight points have been raised:

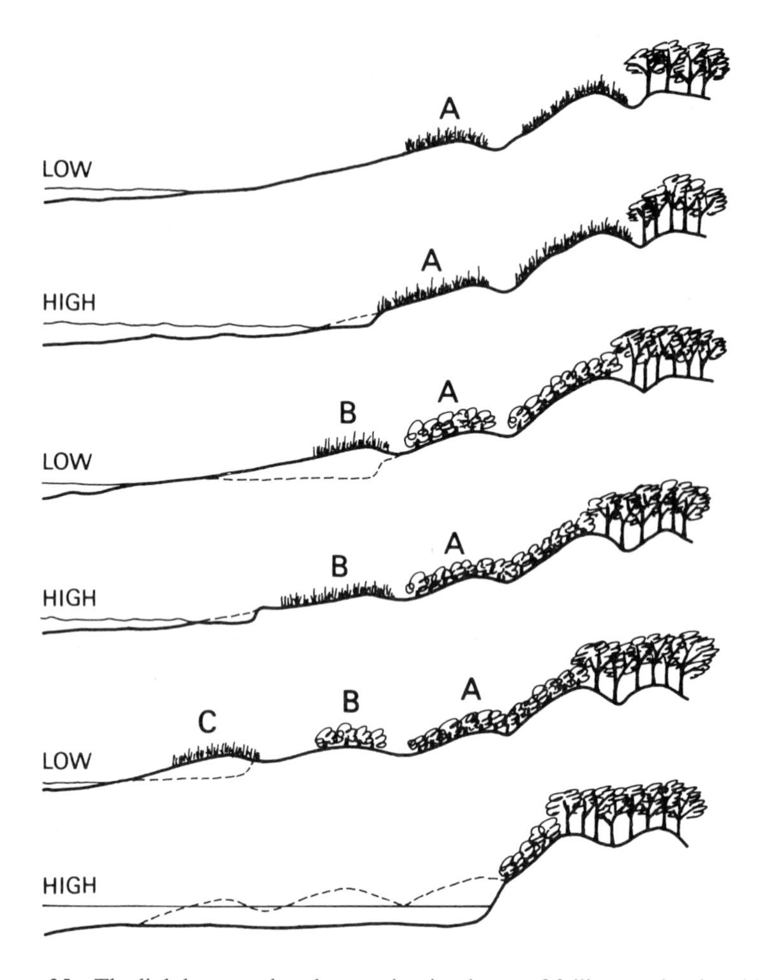

Figure 35 The link between beach accretion in phases of falling sea level and beach erosion in phases of rising sea level is well documented on the shores of Lake Michigan, North America, where a succession of dunes (A, B, C) formed during episodes of emergence was eventually trimmed away by wave action when the lake level rose (after Olson 1958)

1. A pre-condition of the Bruun Rule is that the beaches subjected to a sea level rise were initially in equilibrium (Pilkey et al. 1993). This has been interpreted in various ways (p. 55), but it should be noted that beach erosion is already widespread, so that only a small proportion of the world's sandy beaches are presently in equilibrium.

2. It is necessary to determine a seaward boundary for the application of the Bruun Rule (i.e. the extent offshore of the profile that will be

restored at a higher level). Bruun (1990) suggested that the boundary should be at the line where predominantly coarser nearshore sediment gives place to generally finer offshore material, where the water has become too deep for waves to move material of beach calibre. This requires detailed sedimentological surveys of the sea floor, which were not necessarily available before a sea level rise began, and it is often difficult to distinguish nearshore from offshore sediments. On many coasts there are gradual transitions, and commonly the nearshore topography is variable, with sand bars, rocky outcrops, biogenic structures such as coral reefs, or a muddy substrate offshore.

3. Restoration of the transverse profile can only be completed after the sea has become stable at a higher level, coastline recession coming to an end as a new equilibrium is attained. The Bruun Rule is likely to underestimate the extent of beach recession while sea level is actually rising, and is difficult to apply to a prediction of an accelerating sea level rise without any indication of the level at which it will eventually stabilise.

4. The Bruun Rule deals only with sediment interchange between the beach face and the nearshore sea floor, and takes no account of gains or losses as the result of longshore drifting and their effects on changing beach profiles. It is thus more applicable to swash-dominated than to drift-dominated beaches. Many beaches also lose sediment blown to backshore dunes, washed into tidal inlets, or swept over barriers or spits into lagoons and swamps. Others are augmented by sediment washed down coastal slopes or blown from hinterland dunes. These inputs and outputs can modify beach profiles: for example, an abundant sediment supply from alongshore can maintain or even prograde a beach during a phase of rising sea level, either directly, or by shallowing the nearshore zone so that shoreward drifting ensues.

5. Where the transverse beach and nearshore gradient is low, as on parts of the Caspian coast (Kaplin 1989), the effects of a rising sea level are complicated because storm waves steepen and re-shape the upper beach profile by throwing sand or gravel up above high tide level to form a barrier in front of a low-lying area which becomes submerged as a lagoon. The barrier and lagoon are then driven landward as sea level continues to rise. As it does so, the steepened beach on the seaward side may maintain its profile as sediment is lost, partly by landward overwash and partly by seaward transfer (Ignatov et al. 1993).

6. There appear to be variations in the applicability of the Bruun Rule in differing wave energy environments. On the generally low to moderate

wave energy coasts of New Jersey and Maryland, Everts (1985) found that the Bruun Rule overestimated measured beach recession, but on the Pacific coast, where wave energy is greater, erosion was between 2 and 4 times as much as predicted.

7. The Bruun Rule takes no account of changes in processes or rates of sediment supply and removal resulting from the climatic variations that may accompany a sea level rise, such as increased rainfall and runoff, stronger winds and stormier seas.

8. Where beaches are bordered by nearshore sand bars a sea level rise may be accompanied by upward growth and landward movement of these bars as the beach is cut back, providing there is a sufficient supply of suitable sediment available to maintain this profile. The Bruun Rule could be extended to predict that the landward retreat of the outer limit of the sand bars will be equal to the extent of beach recession, so that the overall profile is preserved. A complication is that sand bars usually consist of finer material than is present on the beach face, so that erosion and seaward movement of beach sand may not provide sediment of suitable calibre for their maintenance, at least in their existing form. On the shores of Lake Michigan, Dubois (1977) found that beach-face erosion as lake level rose was matched by accretion on the landward side of the nearshore bar, which widens as its outer slope remains unchanged.

A rising sea level can certainly be a cause of beach erosion, and is likely to become the dominant cause if the prediction of a world-wide sea level rise, due to the augmented Greenhouse Effect (discussed on p. 107), proves correct (Bird 1993b). The sequence envisaged in the Bruun hypothesis has probably contributed to the erosion of beaches where sea level is already rising (e.g. on subsiding coasts), along with the other factors listed in Table 3.

Geological evidence from past marine transgressions

The prediction that beach erosion will be initiated or accelerated by a rising sea level has to be reconciled with evidence that in the geological past some marine transgressions were accompanied by shoreward drifting of sea floor sediment, and that beaches formed and prograded on coastlines as sea level rise slackened and came to an end. This was the case on many coastlines during the later stages of the Late Quaternary marine transgression (pp. 4, 12), which brought the sea up to, or close to, its present level about 6000 years ago. Some of the beaches that were then formed have since prograded as Holocene beach ridge plains and barriers, although where sea level is still rising (as on the Gulf and Atlantic coasts of the United States), many beaches are now the seaward fringes of barriers transgressing intermittently landward.

It should be noted, however, that the Late Quaternary marine transgression advanced across a land surface that had previously emerged from beneath the sea during a marine regression that occurred about 80 000 years ago. This surface had been strewn with beach deposits left stranded as the emergence took place, and to these were added fluvial and aeolian deposits, as well as glacial and periglacial deposits in high latitudes. There were also unconsolidated materials where shelf rock outcrops had been subaerially weathered during the prolonged low sea level phase. The Late Quaternary submergence thus took place across a shoaly topography, from which sand and gravel deposits were derived and swept landward to form beaches and barriers, many of which subsequently prograded as the marine transgression came to an end. There was, of course, extensive recession of the coastline as the Late Quaternary marine transgression proceeded and the continental shelves became submerged, but as sediment was swept shoreward progradation did occur on many beaches and barriers in its later stages, and particularly where there has been an ensuing stillstand of sea level bordering a stable coast.

With rare and localised exceptions, beach erosion is now prevalent on coasts where sea level has been rising in recent decades, notably where coastal land subsidence is taking place: beaches are being reduced and barriers narrowed or driven landward, and in general sediment is moving seaward to the continental shelf rather than shoreward (Bird 1993b). The present sea floor topography and sediments are very different from their equivalents when the Late Quaternary marine transgression began, because the existing sea floor profile has become adjusted during the Late Quaternary marine transgression and its aftermath. It is now typically gently inclined or concave seaward, shoaly areas being exceptional, and shoreward drifting highly localised. Beaches on submerging coasts are generally in retreat.

If a further sea level rise occurs, as predicted, the possibility of beach erosion being diminished or offset by shoreward drifting of sea floor sediment is likely to be confined to parts of the coast where nearshore areas are supplied with a superabundance of sediment, mainly from rivers. Some deltaic beaches may thus be built upward and seaward as sea level rises, assuming that hinterland levels are maintained by alluvial aggradation, but it is difficult to find evidence of this on existing deltas where sea level has recently been rising. Some examples are found bordering the deltas on the north coast of Java, where beaches have been maintained or prograded close to the mouths of rivers yielding large quantities of sediment during recurrent floods (Bird and Ongkosongo 1980). Beaches could also be maintained or prograded during a phase of rising sea level where there is a sufficient longshore sediment supply from the erosion of beaches and cliffs along the coast. Transgressive barriers of the kind seen on some submerging coasts are likely to be overwashed, submerged and dispersed by an accelerating sea level rise. In the geological record as a whole beaches are poorly represented, having been transient features, preserved only in association with identifiable coastal deposits that survived succeeding marine transgressions and regressions (Davies et al. 1971).

It is therefore unlikely that many existing beaches will be maintained or prograded during a further phase of rising sea level, past stratigraphic indications of their evolution during marine transgressions being of little relevance to present and predicted conditions. If sea level rises across existing coastal plains, especially those which carry extensive dune systems, there may eventually be the formation and progradation of beaches by shoreward drifting of sediment up to the new coastline on the contour at which submergence comes to an end, but by then present-day beaches will have been submerged, buried or destroyed. The short-term response to a rising sea level will thus be the onset or acceleration of erosion on existing beaches, except where this is countered by a sufficient and sustained nearshore or longshore supply of suitable sediment (p. 81). With this exception, beaches can only be maintained in their present positions during a period of rising sea level by nourishing them (p. 210), and it will be necessary to prevent flooding of backing lowlands by building sea walls or artificial land fill.

Diminution of fluvial sediment supply

On coasts where beaches have been supplied with sediment carried down to the coast by rivers (p. 6), erosion is likely to follow any reduction in sediment yield to river mouths as a result of reduced runoff (Brownlie and Brown 1978). This may be due to diminished rainfall or less melting of snow and ice within the river catchment, but is more often a sequel to the building of dams to impound water upstream (Fig. 36). These have intercepted the discharge of river sediment, cutting off the supply of sand and gravel to beaches at and near the river mouth. The result is the onset of erosion on beaches that were formerly maintained or prograded by the arrival of this fluvial sediment. Erosion will develop more quickly, and become more severe, on coasts where there is also strong longshore drifting of sediment away from river mouths.

The best-known example of such erosion is on the shores of the Nile delta, where sandy beaches had been prograding for many centuries as the result of the longshore spread of sandy sediment delivered to the mouths of Nile distributaries. Erosion of beaches near the mouths of the Rosetta and Damietta distributaries was first noticed early in the present century, soon after barrage construction began upstream in 1902. It became much more rapid and extensive after the completion of the Aswan High Dam in 1964,

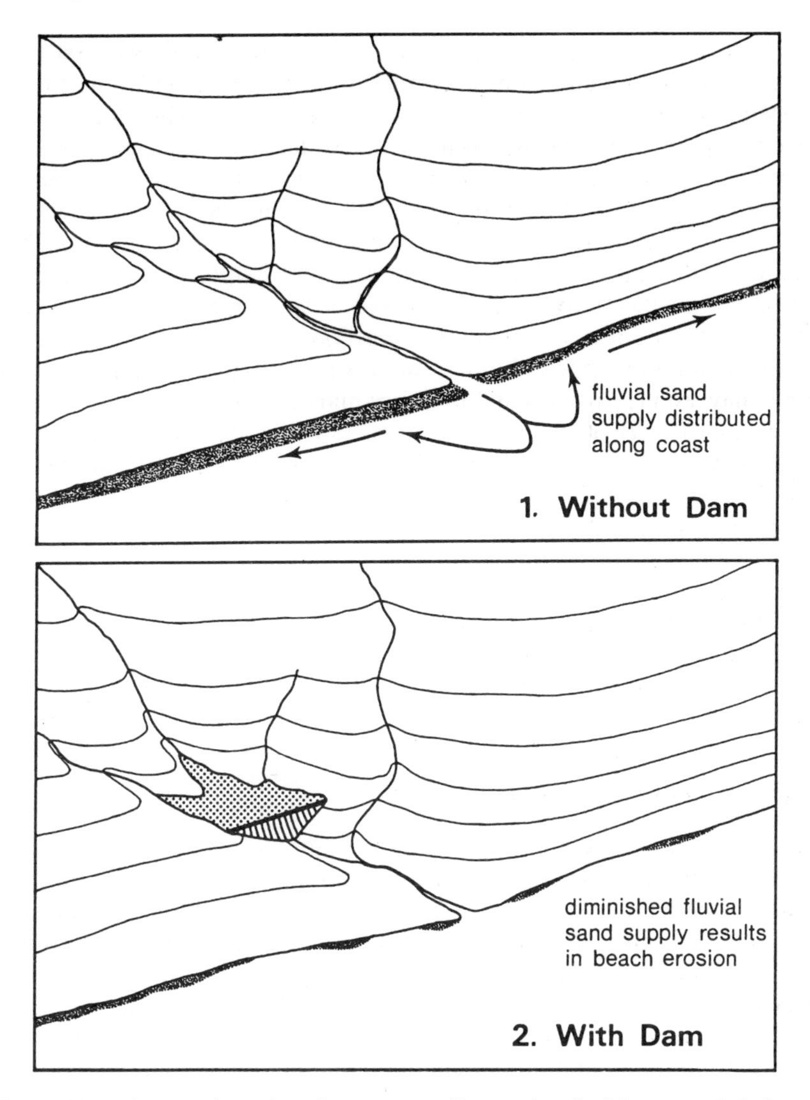

Figure 36 Where a river dam intercepts sediment that had been carried downstream and distributed to beaches along the coast (see Fig. 3) the outcome of the diminution in sediment is generally the onset of beach erosion

which resulted in large-scale sediment entrapment in the Lake Nasser reservoir, and during the next few years, beach erosion on parts of the deltaic coastline attained annual rates of up to 40 m (Orlova and Zenkovich 1974). Some of the sediment removed from these beaches drifted eastward along the coast towards Port Said, but much has been lost offshore (Lofty and Frihy 1993).

A similar sequence of events has occurred on other deltaic coastlines following dam construction upstream. Examples include the Rhône delta in France, the Dnieper and Dniester deltas in the Ukraine, the Citarum delta in Indonesia, and the Barron delta in north-eastern Australia. Ly (1980) documented beach erosion on the coast of Ghana as a sequel to the building of the Akosombo Dam on the Volta River. On the coast of southern California, where relatively high wave energy has largely suppressed delta development, beaches nourished by longshore distribution of sand and gravel delivered by rivers have become depleted since dam construction reduced fluvial sediment yields (Norris 1964). Barrages built on the Tijuana River have retained about 600 000 m^3 of sediment annually, and this has deprived beaches north of the river mouth of a sediment supply, so that erosion has ensued, particularly on Imperial Beach (Paskoff 1994). Beach erosion on the Caucasian Black Sea coast has been attributed to the damming of rivers (Zenkovich 1973), and in northern Sweden, where most rivers have been harnessed for hydroelectric power production, the necessary dams have so diminished fluvial sediment yields that deltaic coasts such as that of the Indal River are eroding despite continuing isostatic uplift of the land.

Beach erosion near river mouths has also followed the dredging of sand from river channels, as in the Tenryu River in Japan (Koike 1985), and cessation of hinterland mining that had previously augmented the fluvial sediment supply to the beach (p. 9). Examples of this can be seen at Par and Pentewan in south-west England, where such beaches are being reshaped and reduced by erosion and sediment attrition.

Another cause of beach erosion has been the reduction of fluvial sediment yield as a consequence of successful soil conservation works (slope terracing, runoff management, reafforestation) in the hinterland, as exemplified by the rivers draining to the Gulf of Taranto in southern Italy (Zunica 1976). Diminished river flow during prolonged droughts has had similar consequences. In southern California, Orme (1985) found that beach erosion occurred during dry periods, especially in 1939–1968, and that beaches were restored during wet years such as 1969, 1978, 1980 and 1983, when the fluvial sediment supply revived. In 1995 beach erosion was increasing on the coast north of the mouths of rivers in north-east Queensland, Australia, particularly the Burdekin, as a sequel to a 5-year drought. Previously these rivers had delivered substantial amounts of sand to their mouths during floods, and this was reworked and drifted northward by the prevailing south-easterly waves to build beaches and spits (p. 39). After several years of drought and diminished fluvial runoff the supply of sand to the river mouths has not been maintained, and beaches close to the river mouths are eroding as south-easterly waves continue to move sand away northward.

The importance of a river in supplying sediment to a beach is also illustrated where there have been natural or artificial diversions of the river mouth. Beach erosion has occurred around outlets from rivers that have been diverted to other parts of the coastline, where a new delta may begin to form. Natural diversion of a river mouth during major floods occurred on the Huanghe in China in 1852 (Chen Jiyu et al. 1985) and on the Medjerda delta in Tunisia in 1973 (Paskoff 1994). Artificial diversion (i.e. the cutting of canals to a new outlet) took place on the Po in Italy in the 16th century (Fabbri 1985) and on the Brazos River in Texas in 1929 (Shepard and Wanless 1971), and has occurred on several deltas in Indonesia. In each case there has been erosion of beaches formerly deposited at and near the abandoned outlet.

On the coasts of Greece and Turkey beach erosion has been attributed to the diminution of sediment yield from rivers resulting from long-continued soil erosion in the river catchments, where increasing areas have been stripped of all unconsolidated material, leaving bare rock widely exposed. An example of this has also been documented from the Ligurian coast in Italy (p. 9).

On the world scale, much beach erosion has been caused by diminished fluvial sediment yields, but there has also been much erosion on beaches remote from river mouths, and for this some other explanation must be sought.

Reduction in sediment supply from cliffs

The supply of sediment from erosion of cliffs and foreshore rock outcrops may diminish for various reasons. Cliff erosion is often preceded and accompanied by subaerial weathering, and runoff after heavy rainfall may cut ravines or gullies in the cliff face. This has been described from Black Rock Point, on the shores of Port Phillip Bay, Australia (Bird 1993a). Sediment slumps or is washed down from the cliff on to adjacent beaches in rainy periods, but if annual rainfall diminishes, or runoff is controlled (e.g. by the insertion of drains along the cliff crest to intercept the water flow and prevent it flowing down the cliff face), the supply of sediment to nearby beaches will diminish, and they will become depleted. Stabilisation of coastal landslides has a similar effect, as has the cessation of coastal quarrying, halting the supply of waste material to the beach. Thus beaches at Porthallow and Porthoustock on the coast of south-west England were augmented by quarry waste between 1878 and 1985 (see Fig. 7), but are now diminishing.

A decline in the strength and frequency of wave attack on cliffs may occur as a result of a climatic change leading to calmer conditions in coastal waters, or when waves are reduced by nearshore shoaling or the growth of reefs. Zenkovich (1967) argued that if sea level remains unchanged, the rate of cliff recession will eventually decline as the result of the widening of the shore platform that is cut in front of them, across which wave action gradually weakens.

The commonest cause of diminished sediment yield to beaches from cliffs is the construction of sea walls along their base. Beach erosion at Bournemouth in England occurred during the progressive extension since 1900 of a promenade on the seafront (Fig. 37). This was intended to halt cliff recession, but it also cut off the sand supply which had come from the cliffs, and with little if any replenishment of the sand and gravel carried away by longshore drifting the Bournemouth beach gradually diminished. On the shores of Port Phillip Bay near Melbourne, Australia, beach erosion is a sequel to the walling and stabilisation of formerly eroding cliffs of Tertiary sandstone (Bird 1993a). Other examples of beaches depleted as the result of the building of sea walls on adjacent cliffed sectors include Ediz Hook, in the state of Washington, USA (Schwartz and Terich 1985), and Byobugaura in Japan (Sunamura 1992).

On the east coast of England, Clayton (1988) found that erosion of 33 km of cliffs in Norfolk, averaging 25 m in height, had under natural conditions yielded 500 000 m^3/year of sediment, about two-thirds of which was sand and gravel supplied to beaches extending for more than 60 km downdrift. With artificial stabilisation of 70% of these cliffs during the past century (many of the coast protection works had proved to be only partially effective in halting cliff erosion), the sediment yield has been reduced by 25–30%, and beach erosion is spreading downdrift.

In situations where cliff outcrops of sandstone or conglomerate, yielding sand or gravel to nearby beaches, are backed by contrasted rock types, such as a massive resistant formation, or soft silts and clays that do not yield beach-forming sediment, cliff recession may eventually expose these. Beaches formerly derived from the sandstone or conglomerate may then no longer be maintained.

Reduction of sand supply from inland dunes

Beaches that have been supplied with sand from dunes spilling on to the shore may start to erode if the sand supply is reduced or terminated because the dunes have become stabilised. This may result from the natural spread of vegetation, or from conservation works such as the planting of grasses or shrubs, the spraying of chemicals such as bitumen or rubber compounds, or paving, road-making and building over the dune surface. Alternatively, the sand supply may run out because the whole of the available dune has moved on to the shore.

Examples of beach erosion resulting from diminished aeolian sand supply have occurred on the south-facing Cape Coast of South Africa, where the prevailing westerly winds are driving dunes along the coast, as at Sundays River. On the Australian coast similar features have been observed. Thus

Figure 37 The coastline at Bournemouth, England, formerly consisted of receding cliffs cut into gravel-capped Tertiary sandstones, and beaches were supplied with sediment derived from these by runoff and marine erosion (above). When a promenade was built, and the cliffs stabilised, the supply of sand and gravel to the beaches was cut off, and erosion ensued (below)

on Phillip Island the partial stabilisation of dunes that had been spilling eastward across the Woolamai isthmus to nourish the beach in Cleeland Bight was followed by beach erosion (Fig. 38), and in South Gippsland stabilisation of dunes drifting eastward across the Yanakie Isthmus led to depletion of beaches on the south-western shores of Corner Inlet (Bird 1993a).

Diminution of sediment supply from the sea floor

There is much evidence, particularly on oceanic coasts, that sea floor sand or gravel has been collected and drifted shoreward by wave action during and since the Late Quaternary marine transgression (Fig. 39) (p. 92). As sea level rose across the continental shelf, waves reworked the sediment that had previously been deposited by rivers or wind action, and eroded material from weathered rock outcrops. The coarser components were swept shoreward, and deposited as sand or gravel beaches. As the marine transgression came to an end, continuing shoreward drifting prograded the beaches, often forming successive backing beach ridges and parallel dunes.

Shoreward movement of sand is still continuing where sediment is being washed in by waves and currents from nearshore shoals, but on many coasts this supply of sediment has declined: as the transverse nearshore profile becomes smooth and concave the beaches can no longer receive sediment from the sea floor. If there is no compensating input of sediment from other sources (such as rivers) progradation comes to an end, and with continued input of wave energy the transverse nearshore profile migrates landward, so that the beaches are progressively consumed. Many beaches and barriers that prograded earlier in Holocene times have passed into this condition, and are now being cut back by erosion (see Fig. 32).

The onset of erosion has come at different times in different places, because the development of the concave profile and the cessation of shoreward drifting of sediment from the sea floor do not occur simultaneously along the coast. Tanner and Stapor (1972) described this change on beaches as a transition from 'an economy of abundance' to 'an economy of scarcity' of sediment supply, and the sequence has been described as a 'maturing of the system'. They illustrated the sequence of events by reference to the erosion that had developed on the beach ridge plain at Cabo Rojo, south-east of Tampico in Mexico. It also provides an explanation for the erosion previously described on the Ninety Mile Beach, in south-eastern Australia, which borders a sandy barrier formerly prograded by accretion of sand supplied from the adjacent floor of Bass Strait, and is now being cut back by marine erosion (see Fig. 33). There is still plenty of sand in the nearshore area, but the transverse profile has become smooth and concave, and there is no longer shoreward drifting of sand to this beach.

Beach erosion could also result from diminished production of shelly

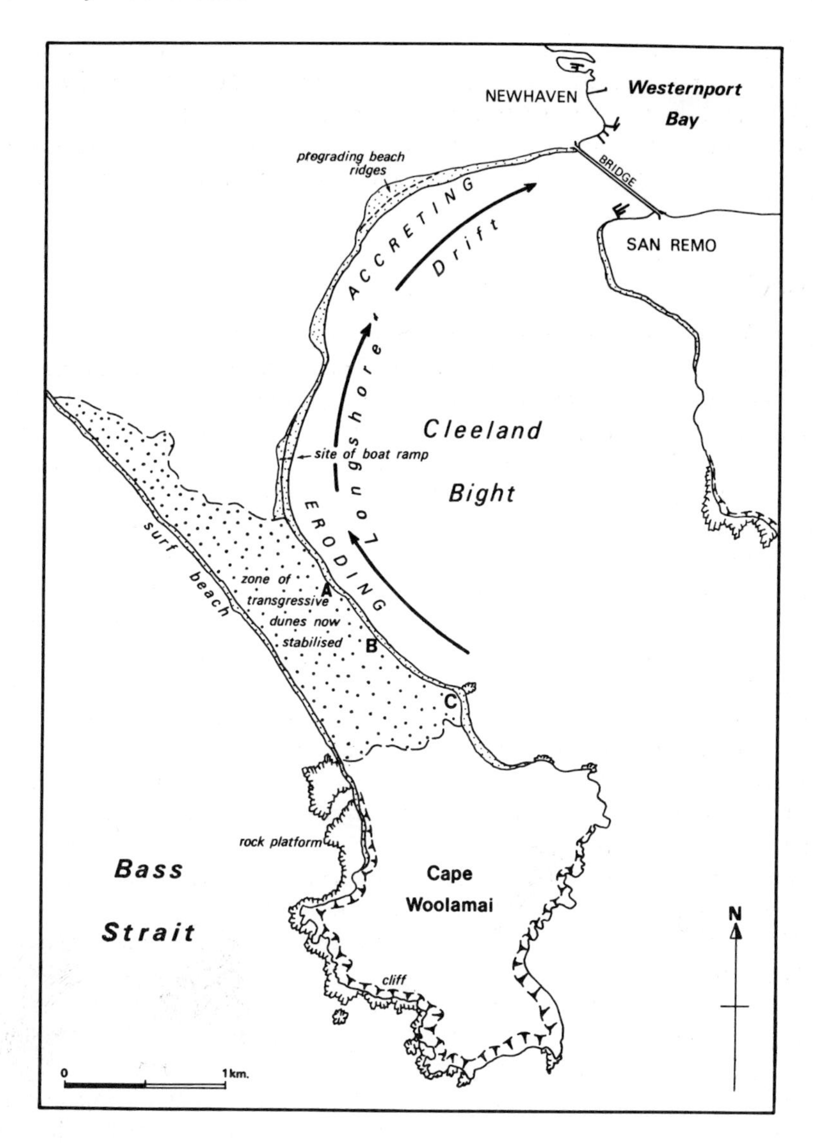

Figure 38 The beach in Cleeland Bight, Phillip Island, subject to northward longshore drifting, was maintained by the supply of wind-blown sand across the isthmus north of Cape Woolamai to the beach at A, B and C (see Fig. 5) until the dunes were stabilised by planting grasses and shrubs. As the wind-blown sand supply diminished, erosion has become prevalent in the southern part of the Bight, and is extending northwards

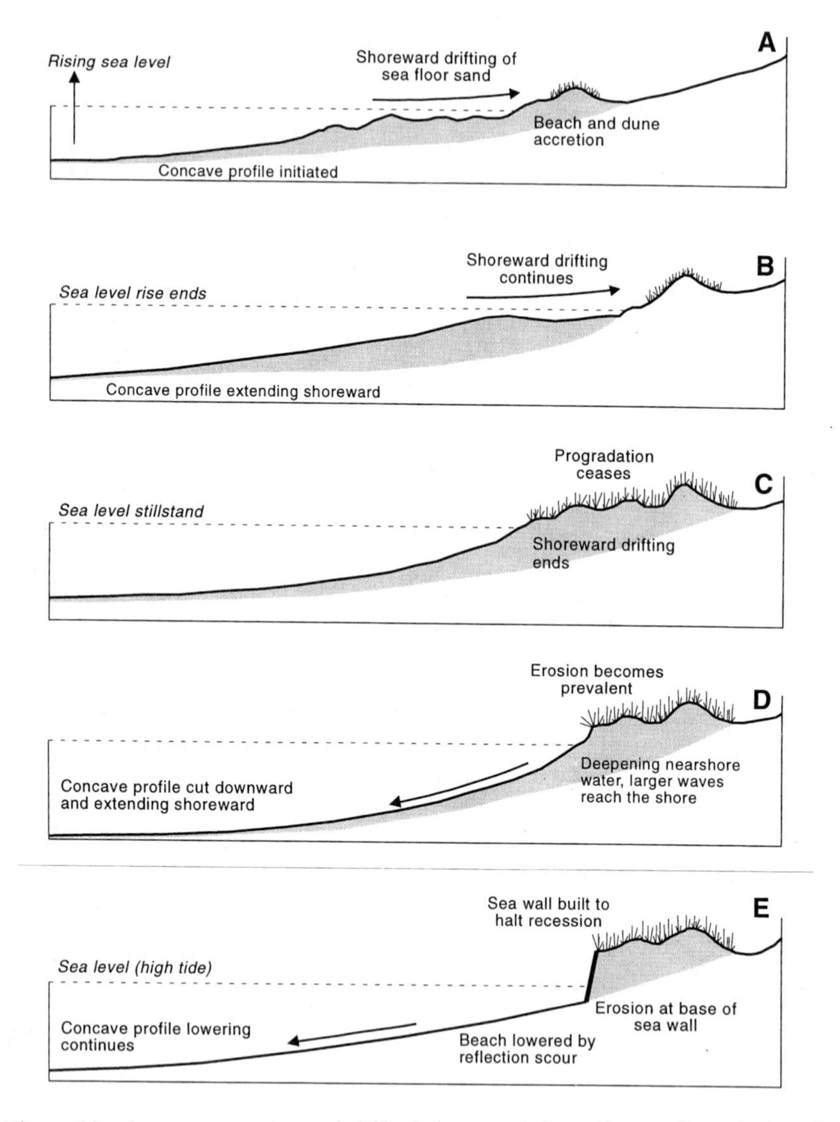

Figure 39 On many coasts sand drifted shoreward from the sea floor during the world-wide Late Quaternary marine transgression (A), and for a time after that transgression came to an end (B), so that beaches and barriers prograded. With the attainment of a smooth concave sea floor profile progradation ceased (C) and subsequently the landward migration of that profile has been accompanied by beach erosion and coastline retreat (D). Building of a sea wall along the eroded margin to halt coastline recession results in further lowering of the beach (E). Compare Fig. 22

Figure 40 Extraction of sand and gravel from the beach at Klim, in northern Denmark, was followed by erosion of the low dunes behind the beach (right)

deposits washed in from the sea floor because of ecological changes, such as the destruction of shell fauna by pollution. It is possible that this has reduced shelly beaches fringing the shores of Corio Bay, near Geelong, Australia. The supply of sand and shingle from the sea floor has diminished where the growth of seagrasses or other marine vegetation has impeded shoreward drifting and trapped the sediment offshore.

Extraction of sand and shingle from the beach

Where sand or gravel have been removed from a beach for use in road making and various building operations the shore profile has been artificially lowered, allowing larger waves to attack the beach more strongly during storms. This has increased beach erosion, for example at Klim, on the north coast of Jutland, Denmark (Fig. 40), and there has been similar depletion of beaches on the south coast of England, where pebbles have been quarried at West Bay on the Dorset coast for use in industrial filters, and from nearby Seatown for ornamental purposes. Gravel has also been taken from the beach at Gunwalloe on the south coast of Cornwall. Beaches in Jersey and Guernsey were much reduced by the extraction of sand from beaches by the German occupying forces during the second world war for use in building bunkers and gun emplacements, and increased backshore erosion has led to the construction of massive sea walls. Another example of beach and backshore erosion resulting from sand extraction has been reported from Casuarina Beach, near Darwin in northern Australia (Gowlland-Lewis et al. 1996).

Beaches have been depleted by the extraction of shell sand and gravel for agricultural use in Cornwall, where this is permitted below high tide level by an Act of Parliament dating from 1609. Such extraction has traditionally been on a small scale (50–100 tons per year) from several beaches, notably at Bude, but recent use of bulldozers to take large quantities from the beach at Poldhu, near Mullion, severely depleted the beach and called this historical right into question. Shelly material is also taken from beaches on islands in the Hebrides (Mather and Ritchie 1977), and at Kinnego Bay in County Donegal (Carter 1988) for aggregate and liming. The assumption seems to be that as the shelly material is of marine origin it will be naturally replenished, but the relative rates of extraction and natural replenishment have not been investigated.

As well as increasing beach and dune erosion, such extractions are likely to accelerate recession of bluffs or cliffs behind the beach. Removal of shingle from the foreshore at Hallsands, in South Devon, from 1897 to 1902 led to more rapid beach erosion and destruction of the village that had been built on a low ledge (an emerged shore platform and raised beach) in front of the cliffs (Fig. 41) (Robinson 1961). Some beaches have been depleted by the extraction of mineral sands, such as rutile, tin or gold (Fig. 42). In general, beach quarrying has led to coastal instability, but the effects depend on the rate of extraction, the size of the beach unit and the interaction with shore processes. In some cases the effects may be almost instantaneous, in others they may take several years to become obvious.

Intensively used beaches at seaside resorts gradually lose sand as it is removed by visitors to the beach, adhering to their skin, clothes or towels, or trapped in their shoes. The quantities are very small, but the losses are cumulative: no one brings sand on to the beach. Pebbles and shells are also carried away by beach visitors. Regular beach cleaning operations, when bulldozers or tractors scrape or sweep seaweed and litter from the beach, also remove sediment, particularly sand (p. 227). Samples taken from heaps of bulldozed seaweed on the beach at Beaumaris, Victoria, Australia, were found to contain up to 20% (dry weight) of sand, and it was estimated that annual beach cleaning was removing up to a 0.5 m^3 of sand from each 100 m^2 of beach surface. As this beach depletion proceeds, areas of the underlying shore platform are being exposed.

Increased wave energy because of deepening of nearshore water

Increased wave attack resulting from the deepening of nearshore water as a shoal migrated away has caused beach erosion on Benacre Ness, on the Suffolk Coast (p. 58), and the same effect was observed on Matakawa

Figure 41 Destruction of the coastal village of Hallsands, in Devonshire, England, followed depletion of the adjacent beach by the extraction of gravel

Figure 42 Extraction of heavy minerals, notably rutile, from beaches and backshore areas on the east coast of Australia, can modify the beach system and may be a cause of erosion

Island, New Zealand, where beach erosion followed the shoreward migration of a tidal channel in the approaches to Tauranga Harbour (Healy 1977). Similar features have been noted at False Cape, Virginia, and in French Guiana and Surinam (Psuty 1985). Deepening of the nearshore zone off Rhode Island, USA, during the 1976 hurricane permitted larger waves to reach the shoreline, accelerating subsequent beach erosion (Fisher 1980).

Nearshore dredging also deepens the water, and allows larger waves to reach the shore. In Botany Bay, Australia, beach erosion accelerated at Brighton-le-Sands after the bay floor was dredged to provide material for the extension of a runway at Sydney International Airport. Nearshore dredging of algae and seaweed around the coasts of Brittany, in France, is thought to have contributed to erosion of beaches and salt marshes in the Baie de la Seine (Cressard and Augris 1982). Deepening of water offshore by dredging has also caused beach erosion at Portobello, near Edinburgh, and the cutting of a trench across the sea floor in St. Ives Bay in Cornwall in 1994 to lay a sewage outfall was followed by erosion on nearby Porthminster Beach. On the Arctic coast of Russia, increased beach and cliff erosion have been attributed to larger waves arriving as the result of nearshore deepening due to down-warping of the nearshore shelf.

Beach erosion at Colombo, Sri Lanka, was partly due to greater exposure to wave attack following the decay of an old reef a short distance offshore. The destruction of nearshore and fringing coral reefs may initially increase the sediment supply to adjacent beaches, so that a phase of progradation ensues, but as the reefs disintegrate the nearshore water deepens, and increasing wave action leads to beach erosion. This has been observed, for example, on the beaches backing decaying fringing coral reefs on the Perhentian Islands off north-east Malaysia.

Interception of longshore drift

Where jetties or breakwaters (p. 60) have been built to stabilise river mouths and lagoon entrances and improve their navigability, or to create boat harbours, longshore drifting of sand or gravel has been intercepted as updrift accretion and there is often beach erosion on the downdrift side.

There are many examples of this. Breakwaters built alongside natural or artificial lagoon entrances on the coast of Florida, as at South Lake Worth, have intercepted sand drifting alongshore from the north to prograde the beach updrift, while to the south beaches deprived of their longshore sediment supply show erosion. At Lagos, in Nigeria, breakwater construction intercepted eastward-drifting sand and resulted in erosion on Victoria Beach, downdrift of the harbour (Usoro 1985). Lighthouse Beach, to the west, prograded rapidly after the breakwater was built in 1912 to maintain a navigable entrance to Lagos Lagoon, and by 1975 the shoreline had

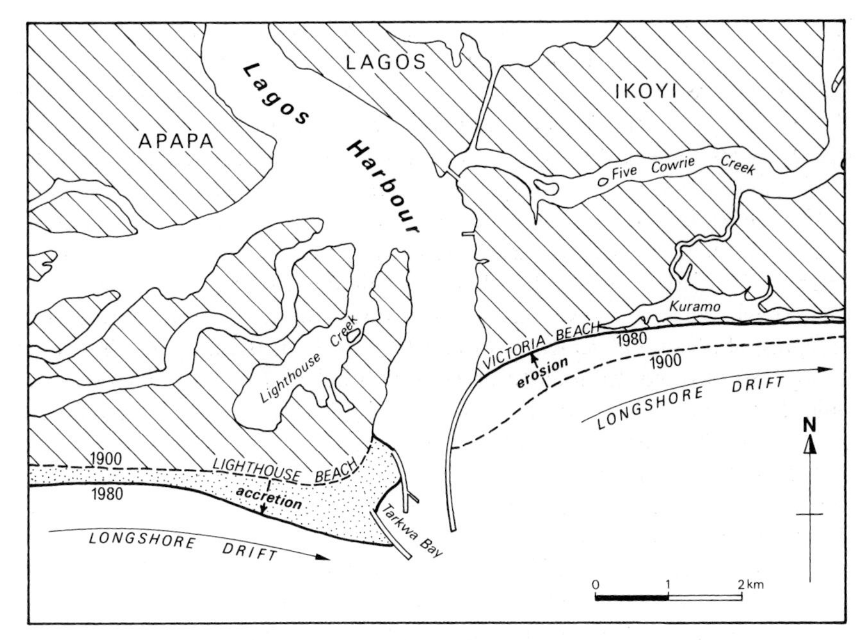

Figure 43 The building of breakwaters at the entrance to Lagos Harbour, Nigeria, has interrupted the eastward longshore drift of sand in such a way as to cause extensive accretion updrift on Lighthouse Beach and erosion downdrift on Victoria Beach

advanced more than 1300 m seaward alongside the breakwater, making a foreland on which successive beach ridges have formed. Victoria Beach had also retreated by up to 1300 m over this period (Fig. 43). Eventually, sand accretion will extend out to the end of the breakwater, whereupon the natural eastward drifting to Victoria Beach will resume, but there will then be problems in maintaining a navigable entrance to the port of Lagos (Ibe et al. 1991).

Jetties built in 1935 to stabilise the inlet south of Ocean City, Maryland, where there is southward drift of beach sand along the Atlantic coast, led to sand accretion on the northern side, while the beach to the south, on Assateague Island, had retreated about 450 m by 1955, the sea breaking through to the lagoon in 1961 (Komar 1976). Similar problems have occurred at Tillamook Bay, Oregon, where a breakwater built north of the inlet has trapped southward-drifting sand, and erosion has become severe downdrift, on Bayocean Spit (Terich and Komar 1974), while on the north-east coast of Malaysia construction of river-mouth breakwaters at Pulau Gaja led to updrift accretion and rapid erosion of the beach downdrift. This erosion is spreading northward and will in due course extend to Pantai Cinta Behari (translated as the Beach of Passionate Love), near Khota Baru.

Figure 44 View eastward towards Newhaven Harbour, south-east England, showing interception of eastward drifting shingle to widen the beach in front of the cliffs

On the south coast of England, where shingle beaches are drifting eastward along the shore, construction of a breakwater to shelter the cross-Channel ferry port at Newhaven in 1845 was followed by updrift accretion (Fig. 44) and increased storm damage and erosion of the beaches downdrift at Seaford, to the east (Fig. 45). Other examples of erosion downdrift from harbour breakwaters include Durban in South Africa, Madras Harbour in India and Santa Barbara in California.

Protruding zones of coastal reclamation can have a similar effect. At Map Ta Phut, near Rayong on the coast of Thailand, a wide protrusion of land reclaimed for port and industrial development has led to updrift accretion of beach sand and erosion downdrift, the proposed solution being further reclamation in front of the eroding beach. Interception of drifting beach material by groynes can have a similar effect. The shingle beaches east of Brighton, England, became depleted after groynes were built to retain a beach for this seaside resort.

Reference has been made (p. 63) to the accumulation of sand supplied by longshore drifting in the harbour at Sandringham, with consequent depletion of beaches at nearby Hampton, on the shores of Port Phillip Bay (Fig. 46).

Interruption of the longshore supply of sand and shingle can also result from the growth of a fringing coral reef or some other depositional feature. Four Mile Beach, in north-east Queensland, has been a relict beach since a

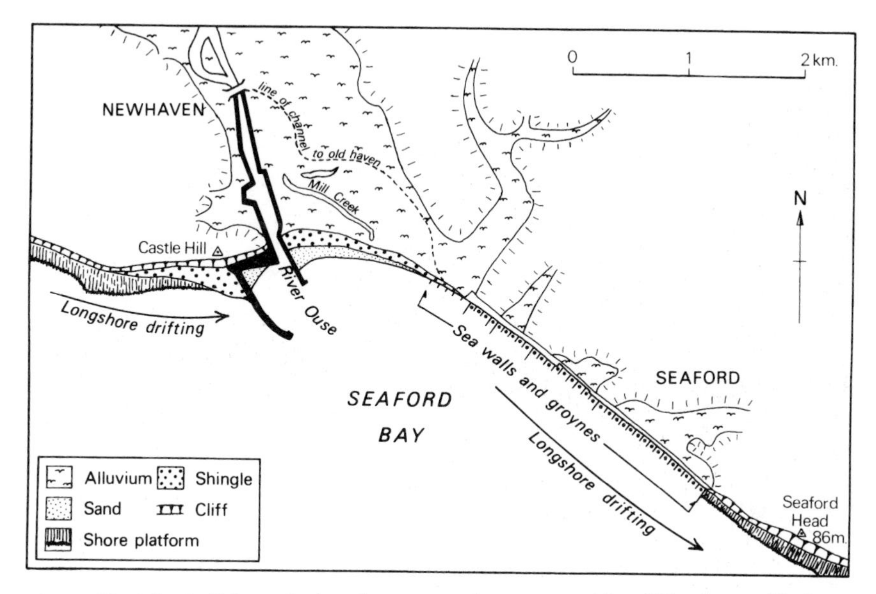

Figure 45 The building of a breakwater on the western side of Newhaven Harbour, in south-east England, has resulted in interception of eastward drifting shingle (see Fig. 44) and depletion of beaches on the coast downdrift at Seaford, where erosion has been countered first by building sea walls and groynes (see Fig. 64) then by beach nourishment (see Fig. 89)

fringing reef developed, cutting off its supply of sediment from the Mowbray River. At Wewak, on the north coast of New Guinea, land uplift led to the emergence of fringing reefs as promontories which segregated a former beach into a series of bays, and cut off the former supply of eastward-drifting sand derived from rivers draining to the coast west of Cape Wom. Thus deprived, the bay beaches have eroded, and attrition (p. 38) has since reduced the shore sediments to fine sand (Bird 1981).

A change in the angle of incidence of waves

Beach erosion can be initiated by a change in the angle of approach of the dominant waves, either because of breakwater construction or because of growth of nearshore reefs or islands or the formation or removal of shoals. This change from a swash-dominated beach alignment to a drift-dominated beach alignment has accelerated longshore drifting. At Kunduchi, Tanzania, the sandy coastline had prograded under the dominance of easterly swell, but the growth of reefs and shoals reduced the effectiveness of these ocean waves, and allowed locally generated and otherwise sub-dominant south-easterly waves to supervene. This resulted in the onset of severe erosion (Bird 1985a). A similar change in wave incidence following

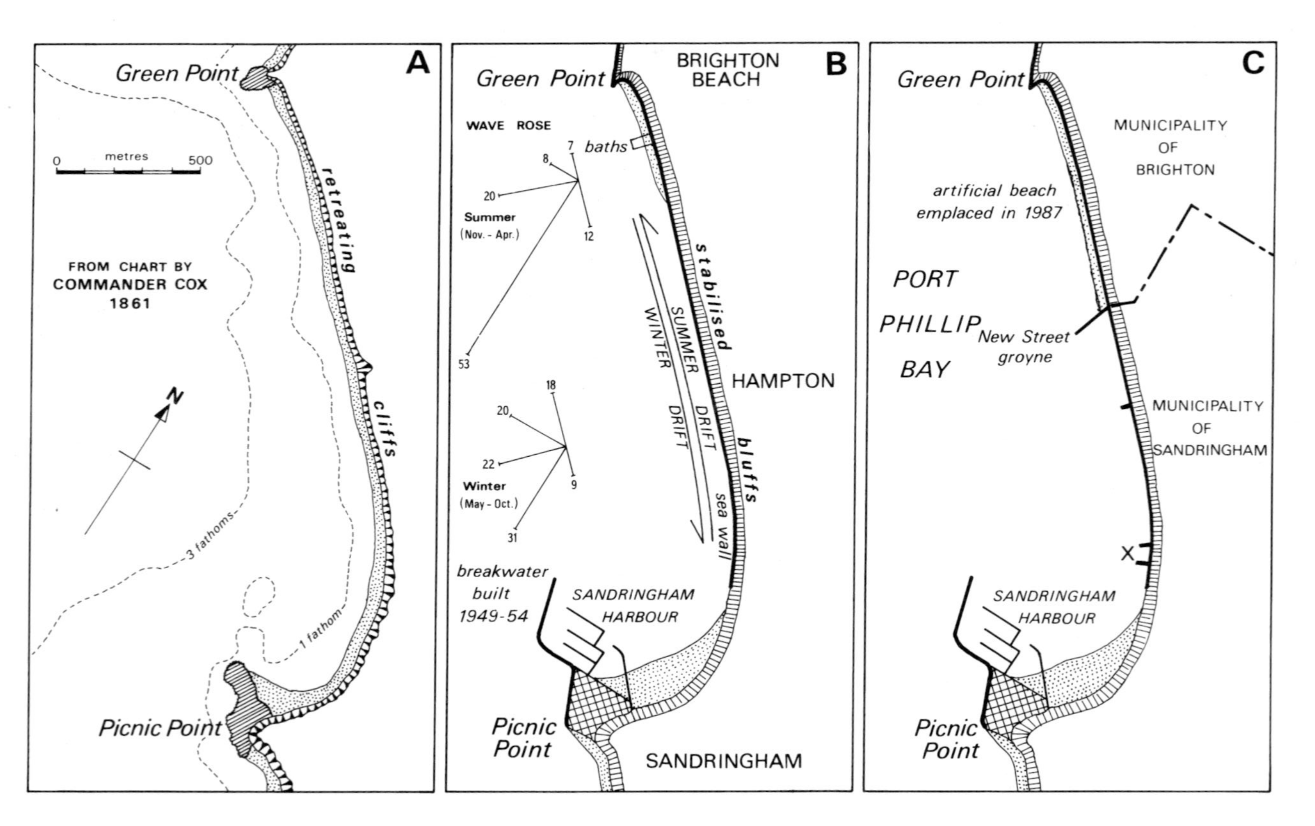

Figure 46 *For caption see opposite*

the construction of the Portland Harbour breakwaters in Victoria, Australia, may have contributed to the onset of beach erosion at adjacent Dutton Way (Bird 1993a). Extension of the harbour breakwater at Albissola on the Ligurian coast of Italy modified the angle of wave approach in such a way that the adjacent sandy beach at Albissola Marina was re-shaped, the eastern part being eroded as the western part widened by accretion (Piccazzo et al. 1992).

Beaches in Marion Bay, Foul Bay and Sturt Bay on the south coast of Yorke Peninsula, South Australia, have been shaped by refracted south-westerly swell, but in recent years all three have shown erosion near the middle and accretion towards their western ends. There being no migrating shoals or artificial structures, this re-shaping may be due to a phase when there have been more south-easterly wind-generated waves than usual.

Intensification of obliquely incident wave attack

Wave attack on one sector of a beach may intensify as a result of the lowering of the beach profile on an adjacent sector, allowing stronger waves to arrive obliquely, and thus accelerate beach erosion. This often occurs after sea wall construction, and the lowering of the beach profile by reflected storm waves (see below; p. 105), which allows larger oblique waves to attack the neighbouring coastline. Nearshore deepening by dredging can have a similar effect. The outcome is that beach erosion spreads alongshore, and if sea walls are extended laterally to counter this, a 'domino effect' of cumulative beach erosion is produced, with each new sector of sea wall on a set-back alignment. This has happened at Point Lonsdale and in Portland Bay, Australia (Bird 1993a).

Increased losses of beach sediment to the backshore

Losses of sediment from the beach to the hinterland occur when onshore winds blow sand inland (Fig. 47), or when storm surges wash beach sand and gravel on to the backshore, or over into lagoons, swales or swamps. Some of the sediment is carried alongshore by winds, waves and currents, to be swept into lagoon entrances or river mouths. If these losses are not

Figure 46 (*opposite*) Evolution of coastal features at Hampton, Port Phillip Bay (location of Fig. 77). The natural coastline had a beach fringe which was maintained with alternations of northward drifting in summer and southward drifting in winter (A). When a breakwater was built to shelter a harbour at the southern end, sand moving southward in winter was trapped (B), so that Hampton Beach was depleted and accretion reduced the area of Sandringham Harbour (see Fig. 24). In 1987 the beach was nourished north of a groyne at New Street (see Fig. 95)

Figure 47 Losses of sand from a beach landward into drifting dunes on Encounter Bay, South Australia. The beach is dissipative, with waves losing energy as they break across a shallow nearshore zone (see p. 51)

compensated by the arrival of fresh supplies of beach sediment, erosion ensues.

In South Australia, beach erosion accelerated after landward movement of backshore dunes followed reduction of their vegetation cover by burning, grazing and trampling. The lowered backshore was then cut back more quickly because of the diminished volume of sand to be removed by wave attack. Overwash during storm surges has driven back sandy barrier islands on the Atlantic coast of the United States (Fisher 1980), and shingle barriers at Blakeney Point and Chesil Beach in England (Steers 1964). In each case the coastline has receded, the beach has been driven back, and there has been beach erosion.

Increased storminess

An increase in the frequency and severity of storms in coastal waters may result in the erosion of beaches that were previously stable or prograding. Beach profiles are cut back and steepened by storm waves until they attain a concave form adjusted to the augmented wave energy. A series of storms in quick succession is particularly destructive, because the second and subsequent events occur on beaches already reduced to a concave eroded profile. Worsening beach erosion on the Atlantic coast of the United States in recent decades may be partly due to an increasing frequency of storms, but detailed long-term weather records are necessary to demonstrate that storminess has increased, and it may be difficult to separate this factor from other causes of beach erosion.

Attrition of beach material

On relict beaches, no longer receiving a sediment supply, agitation of the beach by wave action leads to gradual attrition, and so to a reduction in volume of beach sediment. The beach at Slapton Ley, in Devon, consists of well-rounded and well-sorted granules, in contrast to other shingle beaches on the south coast of England which are still receiving flint pebbles of varying size and shape. Erosion of Four Mile Beach, North Queensland, Australia, occurred after the fluvial sand supply was cut off by coral reef growth (p. 98) and the relict beach sediment has been reduced to very fine sand by attrition. This has become compacted and is now firm enough to land an aircraft, drive a bus or car, or ride a bicycle. As sediment calibre is reduced, such beaches are more likely to lose the increasingly fine sediment by winnowing and removal, either landward into backshore dunes or seaward to bars and bottom deposits. The gradual lowering and flattening of the profile of Four Mile Beach has been accompanied by increased penetration by waves, and the onset of erosion along the seaward margin of backshore dunes. On the Hebridean island of Barra, Scotland, attrition has occurred where a shelly beach has been crushed and compacted by vehicles driven along it. As it was lowered and flattened, the upper beach has been eroded by increased wave scour, and the coastline has retreated (Fig. 48).

Beach weathering

Attrition is a form of physical weathering, but the diminution of beach volume can also be caused by chemical weathering, including the decay of ferromagnesian minerals and the dissolving of calcareous beach sand grains in rainwater, stream seepage or sea spray (p. 38). This lowers the beach profile and permits increased wave attack on the backshore.

Increased scour by wave reflection

Waves breaking against a solid structure, such as a sea wall built of concrete, stone blocks, steel sheeting or timber, or a boulder rampart, are reflected, and the seaward currents produced sweep sediment away from the foot of the wall (Fig. 49). This has been observed on many coasts where sea walls have been built behind a beach to halt cliff recession (Kraus and Pilkey 1988, Tait and Griggs 1990). Storm waves at high tide then overwash the beach to splash against the wall, and their reflection causes beach erosion, which can eventually undermine the foundations of the wall.

Figure 48 Backshore erosion on a shelly beach due to heavy fire engine traffic associated with the use of the adjacent lower beach, exposed as the tide falls, by aircraft landing on the Island of Barra in the Hebrides, Scotland

On the Queensland coast in Australia the sandy beach at the resort of Surfers Paradise was depleted by a series of cyclones, and a large boulder wall was built to safeguard the coastal hotels and apartments (Fig. 50). The boulder wall then caused scour by reflecting the waves, which lowered the beach and prevented the natural restoration that had occurred after previous such erosion by cyclones in 1950, before the boulder wall was inserted.

Reflection scour following the construction of sea walls has lowered beaches in Jersey, in the Channel Islands. On the west coast, St. Ouen's Bay had a wide sandy beach, backed by dunes, formed as the result of shoreward drifting of sand during the Late Quaternary marine transgression (see Fig. 39 A–C), but by the 19th century it was eroding as the result of a diminution in sea floor sediment supply (see Fig. 39, D), and the building of a sea wall to halt coastline recession resulted in further lowering and flattening of the beach (see Fig. 39, E).

Migration of beach lobes

Beach lobes form as the result of alternating longshore drifting, or because of intermittent passage of drifting sediment past an obstacle such as a headland, stream mouth or lagoon outlet (p. 59). Where there is a predominance of longshore drifting in one direction, such lobes may move

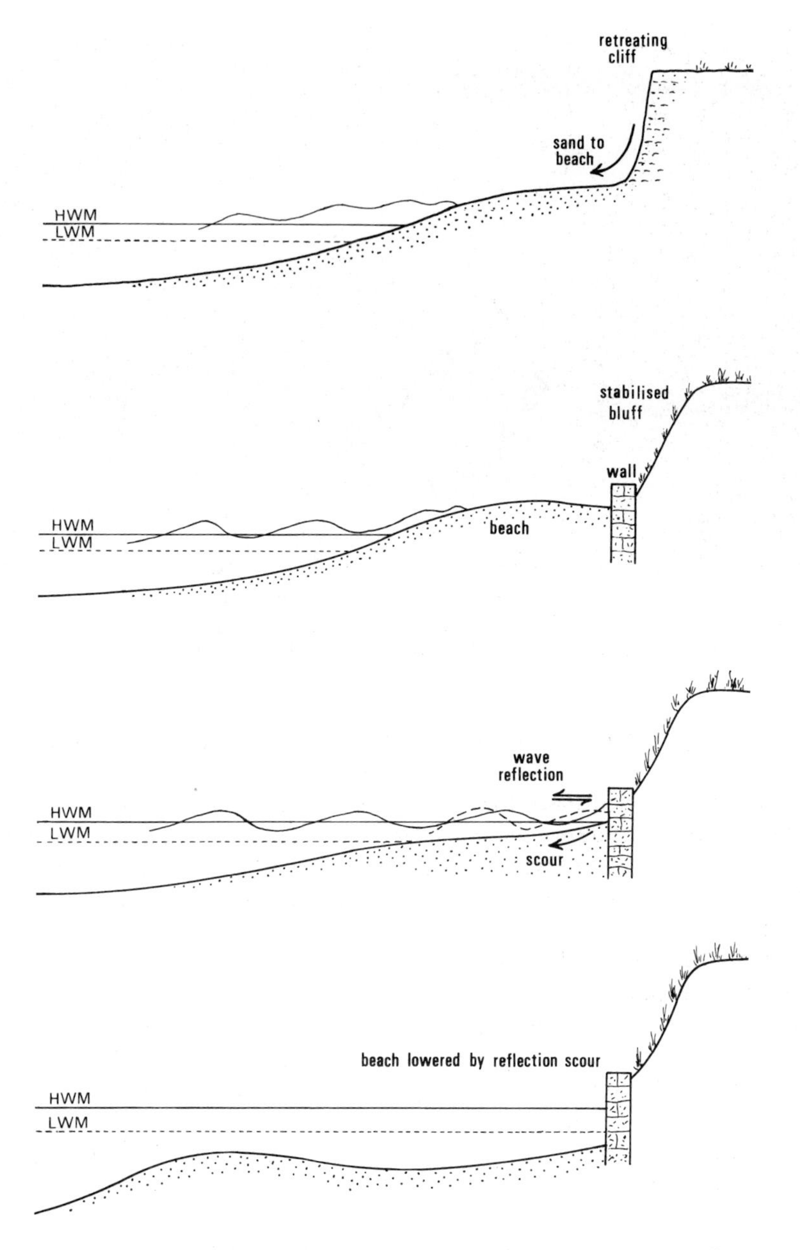

Figure 49 Sequence where a receding cliff, which had been supplying sediment to a beach, is fronted by a sea wall and stabilised (see Fig. 37). Reflection of waves from the sea wall causes beach erosion

Figure 50 Successive cyclones caused severe erosion on the beach at Surfers Paradise in south-east Queensland, resulting in the building of a boulder and rubble wall. This caused reflection scour, preventing natural restoration and further depleting the beach, which has subsequently been artificially nourished and maintained

along the coast, with erosion of the updrift flank and accretion downdrift, usually intermittent, during stormy periods. At a particular point the beach progrades as each lobe arrives, then beach erosion develops as it moves on downdrift. A succession of such beach lobes has been charted, moving alongshore from Somers eastward to Sandy Point in Westernport Bay, Australia (Bird 1985b). Reference has been made (p. 60) to the building of a yacht clubhouse on one of them in the 1970s, and to the problem of it being threatened by beach erosion as the lobe moved on.

A rise in the beach water table

It has long been known that a wet sandy beach is eroded more rapidly by wave action than a dry one. Analysis of changes over 95 years at Stanwell Park Beach, near Sydney, Australia, by Bryant (1985) identified rises in the level of the beach water table as a contributory cause of beach erosion. Such a rise in the beach water table may be due to a rise in sea level, to the ponding or diversion of river or lagoon outlets, to unusually heavy or prolonged rainfall, or to increased discharge following land-use changes in the hinterland. It is marked by extensive seepage, keeping the beach wet at low tide.

Removal of beach material by runoff

Beach erosion can occur as the result of runoff of water during a period of heavy rain, or the melting of snow or ice, either from a backing cliff or steep slope. Beach sediment can be swept into the sea by strong runoff issuing from a stream or drain. These effects are stronger on sandy beaches, especially if they are already wet, than on gravel, where runoff more quickly disappears by percolation. The seepage mentioned in the previous section also contributes to beach erosion by washing sediment seaward as the tide ebbs.

Increased runoff is often due to urbanisation and the construction of roads and other sealed surfaces from which water runs off quickly, instead of percolating into the subsoil as it did before these structures were added.

Diminished tide range

Erosion by waves is more effective where their energy is concentrated at a particular level, rather than dispersed by the rise and fall of a substantial tide. It follows that a diminution of tide range will increase the effectiveness of wave action, and this could initiate or accelerate beach erosion. Examples may be found on the shores of inlets, estuaries or lagoons that become partly or wholly cut off from the open sea by the growth of spits and barriers, or the building of structures such as weirs or barrages, so that tidal ventilation is impeded, or excluded altogether, and wave action is intensified at a particular level. This has been observed on beaches fringing the coastal lagoons at the mouth of the Murray River, in South Australia, which were formerly estuarine and tidal, but were separated from the sea by the construction of barrages in 1940.

Removal of a sea ice fringe

On cold coasts, as in Alaska, northern Canada and Siberia, beaches are protected in the winter by the formation of a fringe of shore ice, and are subject to wave action only during the brief summer thaw. If the climate becomes warmer, and the summer lengthens, waves will reach these shores for a longer period, and beach erosion will increase.

EFFECTS OF A RISING SEA LEVEL

The beach erosion problem will intensify if, as is predicted, a global sea level rise develops in the coming decades as a consequence of man-induced changes in the composition of the atmosphere, leading to the enhancement of the so-called 'Greenhouse Effect', the opaqueness of the atmosphere to reflected solar radiation which maintains existing global temperature regimes. Measurements have shown increasing concentrations of such gases as carbon dioxide, methane and nitrous oxide in the Earth's atmosphere, and these will produce global warming, which in turn will lead to expansion of the oceans and some melting of the world's snowfields, ice sheets and glaciers, resulting in a world-wide rise of sea level (Barth and Titus 1984).

Calculations by the Intergovernmental Panel on Climatic Change in 1990 indicated the probability of a global sea level rise of 15–20 cm by the year 2030, accelerating to 1 m during the coming century, but later predictions suggest a smaller and more gradual rise, of the order of 46 cm by the year 2100 (Wigley and Raper 1993). Changes likely on such submerging coasts have been discussed by numerous authors, including Titus (1986) and Bird (1993b). A sea level rise will generally result in a deepening of nearshore water, so that larger waves break upon the shore, initiating erosion on beaches or accelerating it where it is already in progress. On coasts where the beach fringe is narrow, backed by high ground or marshlands, beaches will soon disappear, but they will persist where they retreat through beach ridge plains or coastal barriers of sand and shingle. Beaches will be maintained where there is a continuing supply of sand and shingle, or where the supply is increased as the result of accelerated nearby cliff erosion, greater sediment yield from rivers because of heavier or more effective rainfall, catchment devegetation, or disturbance by tectonic uplift or volcanic activity. If the nearshore sea floor is built up by accretion, or if the longshore sediment supply is maintained at a sufficient level, beaches may persist, or even be prograded, as sea level continues to rise.

The effects of a sea level rise can already be seen on beaches where land subsidence is in progress, as on the Atlantic seaboard of the United States (Leatherman 1990), in the southern North Sea (Goemans 1986), the northwestern Adriatic, and a number of other parts of the world's coastline. Submergence has already contributed to beach erosion on these coasts, much as predicted by the Bruun Rule, subject to the limitations noted above (p. 80). A global sea level rise will undoubtedly result in beach erosion becoming even more extensive and severe than it is now, and many seaside resorts will lose their beaches. There is a need for planning to take account of such changes, for as coastal development proceeds and intensifies it will become more difficult and more expensive to cope with the effects of a rising sea level (K.M. Clayton 1989).

CONCLUSION

No single explanation can account for the modern prevalence of erosion of the world's beaches, or indeed for the onset or acceleration of erosion on

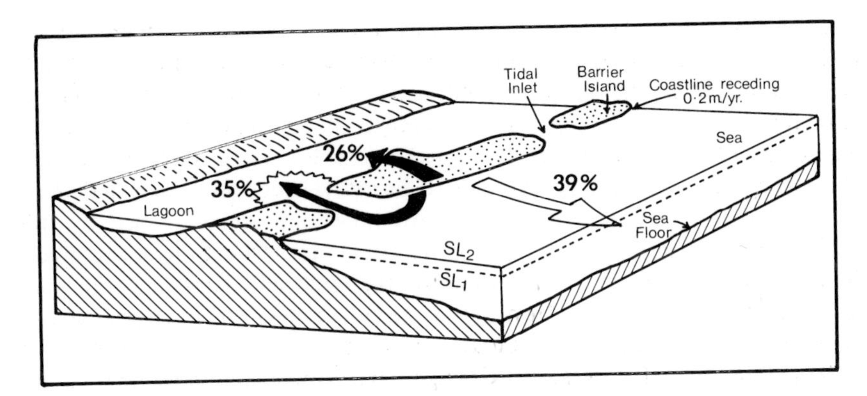

Figure 51 Model of the distribution of sediment losses accompanying recession of sandy barriers in Rhode Island, United States, based on Fisher (1980)

any particular beach. It is not simply the outcome of human activities, artificial structures, a sea level rise, an increase in storminess of coastal waters, or the 'maturing of the system' as the sediment supply from the sea floor dwindled during the Holocene stillstand (p. 12). Each of these factors may have contributed to beach erosion, to an extent which differs from place to place.

An example of multicausality is seen on the Lido di Jesolo, the beach bordering a barrier island fronting the Lagoon of Venice on the Adriatic coast (Zunica 1990). Since this island was developed as a resort in the 1950s, there has been severe beach erosion. The causes include coastal subsidence, augmenting a rising sea level in the Adriatic, nearshore deepening, and changes in current flow and wind and wave regimes, all of which have curtailed sand supply from the sea floor and favoured increasingly energetic wave attack. In addition, reafforestation, reservoir construction and excavation of sand and gravel from river channels have diminished the former sediment yield from the Piave River to this coast. Artificial structures include groynes, which have reduced longshore drifting, and concrete and boulder sea walls, which have increased reflection scour.

The task of ranking the relevant factors and apportioning their contribution to beach erosion on particular coasts requires investigation of past and present patterns and rates of change on beaches and the process systems operating in coastal waters. An example of quantitative assessment of beach erosion was provided by Fisher (1980) from Rhode Island, USA (Fig. 51). He found that between 1939 and 1975 the barrier beach coast of Rhode Island retreated at an average rate of 0.2 m/year, in a period when sea level rise averaged 0.3 cm/year. This was within the range predicted by the Bruun Rule for this rate of sea level rise (0.15–0.3 m/year), but Fisher calculated that 35% of linear beach recession over this period had been due to the washing of sand into tidal inlets and 26% to losses of sand washed or blown over the barrier islands to form migrating dunes and washover fans. This left 15% of the beach retreat accountable as the direct result of submergence, and 24% lost by transference of beach sand seaward. The Bruun Rule thus overestimated the response to sea level rise here. Movement of sand over and between these barrier islands indicates that they were continuing the long-term landward migration, accompanied by the transgression of backing lagoons and marshes, that has characterised much of the Gulf and Atlantic coastline during the Holocene. A similar sequence of evolution is now being observed on the submerging coasts of the Caspian Sea (p. 81).

Responses to beach erosion vary with the nature of the coast and the degree of economic development in a coastal region. On remote and undeveloped coasts beach erosion may be of little importance, and can proceed unhindered: if the eroded sand and gravel are carried alongshore, erosion of one beach may result in accretion on another. Where there are coastal urban or port development or seaside resorts, the initiation or acceleration of beach erosion becomes a matter for concern. The next two chapters examine the various kinds of structural works that have been introduced with the aim of halting coastal erosion and beach loss, and the introduction of beach nourishment to counter erosion.

Chapter Four

Structures for beach protection

STRUCTURAL RESPONSES TO BEACH EROSION

A common response to beach erosion has been to build structures designed to protect and maintain existing beaches or prevent further recession of a coastline. The two aims have sometimes proved incompatible: retreat of a coastline can be halted by building structures such as sea walls or boulder ramparts along the eroding land margin, but this may not achieve retention of beaches. Coastal engineering practice has been much the same in many countries: in the United States it has been guided by successive editions of the Shore Protection Manual produced by the Coastal Engineering Research Center of the United States Army Engineering Waterways Experimental Station (CERC 1987). Here the aim is to provide a brief account of structures used, and their advantages and disadvantages in terms of beach management.

SEA WALLS

A common response to coastline erosion, especially where it threatens to undermine and destroy developed property, such as roads or buildings, has been to construct, extend and elaborate sea walls. Often these are initially banks of earth or other locally available material, particularly if it can be extracted from an adjacent excavation, often a parallel ditch. Subsequent reinforcement may be necessary in response to damage of such banks by storm waves, and these are stages in the evolution of larger, more massive and costly solid structures. Shave (1989) noted such stages in the evolution of the 208 km of sea wall that now line the coast of Kent, in south-east England.

Modern sea walls may be built initially as large, solid structures. Massive stone or concrete walls are designed to withstand the force of the breaking waves, which are reflected seaward, and may scour away the beach. An alternative is to build boulder ramparts, also known as revetments or

Figure 52 Shore protection at Beesands in Devon, south-west England, was by dumping large stone boulders, burying what was left of the shingle beach

riprap, some of which are irregular heaps of rocky debris, others more carefully fitted blocks arranged on a seaward slope, usually on a mattress of sand or gravel (Thorn 1960). There is also emplacement of artificial structures such as tetrapods, made of reinforced concrete, shaped to interlock and remain in position on the shore during phases of strong wave action. Boulder ramparts and tetrapods are less reflective than solid sea walls, and the expectation is that waves will break across them, producing swash and backwash that do not cause erosion on their seaward sides. In some places, rocky debris has been dumped to protect an earlier sea wall from undermining and disintegration by wave attack, or in the hope that a boulder apron would prove less reflective than a solid sea wall, and permit some recovery of a lost or diminished beach (Fig. 52).

Sea walls and boulder ramparts are often introduced to prevent wave attack on the eroding coast, usually a receding cliff, an undermined and slumping bluff, or a truncated dune, in each case fronted by a beach that is insufficiently high and wide to prevent waves reaching the back of the shore. On the north-east coast of Port Phillip Bay, Australia, sea walls built in the 1930s to halt cliff recession resulted in depletion of beaches at Hampton, Sandringham, Black Rock and Mentone, as discussed in Chapter 5.

In addition to damage by the physical impact of waves during storms (Fig. 53), and by scour due to the hurling of sand and gravel against them

Figure 53 Winter storm waves breaking over a sea wall at Black Rock, Port Phillip Bay, Australia, after the beach had been depleted by longshore drifting during the preceding summer (see Fig. 14)

by large waves, sea walls also decay by physical, chemical and biological weathering, a process that can be rapid on certain sandstones and limestones (Clark 1988).

The lateral extent of sea walls and similar structures usually depends on the coastwise length of developed land to be protected, and on such factors as costs and municipal boundaries. Armouring of a particular sector of coast to protect a building or a seaside resort is usually followed by continuing recession of adjacent sectors, so that in due course the protected area becomes a promontory. This has happened at the seaside resort of Mundesley, on the East Anglian coast, which now protrudes between retreating cliffs of soft glacial drift, and at Bray on the dune coast of north-eastern France. Eventually the flanks of such promontories have to be stabilised artificially, and in due course the protected area could become an offshore island.

There are often objections to proposals to build or extend sea walls and similar artificial structures, either on aesthetic grounds (because they look unsightly, even compared with eroding coastal scenery), or because they have adverse ecological effects, such as the loss of habitat for various plants, animals and birds. Sometimes sites of scientific interest, such as cliff exposures of particular geological formations, are concealed from view and made inaccessible for education and research when they are covered by a

Figure 54 On the east coast of Japan concrete tetrapods are constructed on site and dumped on the beach to protect it from erosion during typhoons

sea wall. The 'hard engineering' solution to coastal erosion can look bleak. In Japan long sectors of the cyclone-prone east coast of Honshu are now protected by large sea walls and concrete tetrapods weighing up to 50 tonnes and costing about $US2500 each (Fig. 54). Walker and Mossa (1986) estimated that Japan had about 8000 km of sea walls and tetrapods (about a quarter of the coastline). In Britain there has long been concern at the spread of sea walls and boulder ramparts along the coast. The Royal Commission on Coastal Erosion (1907–11) noted the many coastal structures built in the 19th century, and Sherlock (1922) commented on their lateral expansion, which has continued.

The strongest objection to such structures is that they often lead to the depletion or disappearance of the beach that formerly fronted them (p. 103) (Kraus and Pilkey 1988). Many seaside resorts have seen their beaches depleted as a sequel to the building of sea walls and other solid structures, which cause wave reflection and the scouring away of the beach sediment. As the beaches diminish and the foreshores are lowered these structures are undermined by storm wave attack (see Figs. 49 and 50). Thus the halting of coastline recession has been achieved only at the cost of losing the beach which was the main attraction for seaside holidaymakers. In some places the lowering of the beach has prompted the rebuilding of a new sea wall some distance seaward, with infilling of the intervening area to form a promenade, as at Porthcawl in Wales (Carter 1988).

On the island of Jersey most of the beaches are sandy, and backed by large sea walls built in the 19th century and subsequently repaired and elaborated, and beaches fronting them have been depleted by wave reflection scour so that they are now typically low-lying, submerged at high tide and wet with groundwater seepage at low tide (Fig. 49). The response to the undermining of the base of these walls by wave scour following the lowering of the beach level has been to repair them, sometimes with secondary basal ledges. A review by Hydraulics Research (Wallingford, England) led to the suggestion that the walls be reinforced with basal dumping of boulders or cobble gravel, or the insertion of perched beaches, enclosed and retained by a low wall near high tide level.

Sea walls and similar structures are requested initially as a means of halting coastline recession, and are the engineering response to this problem, rather than for beach protection. If the aim is to retain the beach (and perhaps to use it as a protective feature, reducing or halting coastline erosion), this should be made quite clear before the building of artificial structures is contemplated. The dumping of large boulder ramparts on a shore, usually at the back of a beach to prevent undermining of cliffs by storm wave action, but sometimes to protect an earlier, inadequate wall (see Fig. 52), is often resented by local people and beach users as an alien intrusion, akin to solid sea walls, with the additional disadvantage that boulder ramparts can provide niches for rotting seaweed, litter and rats.

The aesthetic problem may be lessened if the boulders are of rock types that occur naturally on the nearby coast. Although dumping of boulders may be unwelcome, there is a gradation through to cobbles and pebbles which on some coasts may be acceptable as a beach that looks natural, while absorbing enough incident wave energy to protect the coastline. It is more difficult to persuade people to accept a cobble or pebble beach where the natural beach was of sand.

BEACH PROTECTION

Efforts to protect and maintain a beach should be based on an understanding of why it is being depleted, taking account of the 20 possible causes discussed in the previous chapter (see Table 3), and where the sediment is going. Beach depletion by losses alongshore may be reduced or halted by inserting groynes and breakwaters to impede longshore drifting, but where sand and gravel are being lost seaward to the sea floor these will not have much effect, and some kind of underwater trap may need to be considered. Prevention of losses to the backshore, where beach sediment is being washed or blown inland, or across spits and barriers into lagoons or swamps, may require the insertion of solid structures, or perhaps a vegetation baffle, at the back of the beach. Structures have been built to

Figure 55 Boulders dumped on the shore of Sandy Hook, New Jersey, in order to shelter and maintain a sandy beach

protect and shelter a beach (Fig. 55), sometimes by dividing it into compartments (Fig. 56) or actually enclosing it (Fig. 57), but some beaches have been entirely replaced by artificial slopes (Fig. 58).

BREAKWATERS AND GROYNES

A breakwater (p. 60) is a solid structure built out from the coastline, usually to shelter a harbour or protect a river mouth or lagoon entrance. It may also afford additional protection to beaches in its lee, but there are complications on coasts where there is longshore drifting of beach sediment. There is usually progradation updrift of the breakwater and consequent erosion downdrift, as in the several examples cited on p. 98. The ancient stone breakwater known as The Cobb at Lyme Regis, on the south coast of England, shelters a harbour to the west of the town and has reduced storm wave impact on the town beaches to the east, but it has also interrupted longshore drifting of shingle, so that Monmouth Beach to the west has prograded and the town beaches downdrift have been gradually depleted. If the harbour is on the updrift side of the breakwater, longshore drift will be intercepted as accretion within the harbour area, as in Sandringham Harbour, Port Phillip Bay, Australia (Fig. 24, p. 63).

A breakwater built with the intention of intercepting drifting sediment to

Figure 56 The use of pinewood fencing to shelter and retain a beach at Lepe on the shores of The Solent in southern England

Figure 57 In order to preserve the narrow shingle beach at Minehead in south-west England a protective structure was built with sheet iron roofing. The beach was retained, but its recreational value was reduced

Figure 58 In response to recurrent storm surge flooding and erosion the beach on the inner shore of Hel Spit, Poland, has been converted to an artificial slope with a network of concrete forming enclosures in which vegetation has been planted. The coastline has been stabilised, but there is no longer a beach

widen a beach is generally known as an anchor groyne (spelt groin in the United States). Updrift accretion generally results in the beach becoming wider and higher for some distance along the shore, the prograded sector becoming triangular or cuspate and extending towards the outer end of the groyne. The extent and form of updrift progradation depends on a number of factors, including the length and height of the groyne, the availability of drifting beach sediment, its texture and sorting, and the direction and strength of waves and currents, as influenced by the changing nearshore topography. The prograded beach is generally built as a terrace to the level of the swash limit just above normal high tides, but it may be diversified by storm-piled ridges of sand or gravel or by dunes built of sand winnowed from the foreshore. These features either remain mobile, or become fixed by colonising vegetation.

Multiple groynes (groyne fields) have been built on some coasts, especially at seaside resorts such as Felixstowe in East Anglia (Fig. 59), with the aim of retaining drifting beach material and so protecting the coastline (Fig. 60). Beach sand and gravel are intercepted in the intervening compartments, and accumulate until sediment spills over or round each groyne. Often a larger terminal groyne is built at the downdrift end (Fig. 61), or harbour breakwaters may serve this purpose (Fig. 62). As sediment drifting along the shore is trapped by the groynes the supply to

Figure 59 Concrete groynes retain a beach for the seaside resort of Felixstowe, Suffolk, England

downdrift beaches is reduced, and erosion ensues. Erosion is thus transferred along the coast, and there is then a temptation to extend the groyne field. As a result there are sectors of the coastline of England and Wales that now have multiple groynes for several miles. Although there may be some bypassing of beach sediment along the sea floor seaward of groynes, erosion of beaches downdrift is likely to continue until there has been sufficient accretion between the groynes for longshore drifting to resume past them.

Groynes may be built of timber, masonry, sheet metal, boulders or concrete. On the German North Sea island of Sylt, wooden groynes were first used in 1865; they were replaced first by concrete in 1913, then iron in 1927 (Kelletat 1992). Some groynes are solid, impermeable structures intended to prevent longshore drifting of beach material until the beach has been built up to spill over them or round their outer ends. Others are semi-permeable or permeable, usually wide rubble-mound groynes of low profile, designed to intercept and capture a proportion of the beach material drifting along the shore, and allow some nourishment of downdrift beaches to continue.

The optimum length of groynes depends partly on the angle of approach of the dominant waves to the shoreline. Wave crests arriving at 40° to 50° to the shoreline are most effective in generating longshore drifting, and groynes must be relatively long if they are to intercept sediment in the

Figure 60 Successful use of groynes on the southern shore of Botany Bay, NSW, Australia, to retain sandy beaches in the intervening compartments

intervening compartments: the groynes can be shorter where the angle of incidence of the waves is greater or less than this. Groyne length also depends on the nature of the beach material. Groynes can be shorter for shingle beaches, where much of the longshore drifting is caused by the swash of waves breaking on the beach face, but sandy beaches are wider and flatter in profile, especially where they are dissipative (p. 27), and longer groynes are necessary to retain them.

The optimum spacing of groynes (i.e. the width of intervening compartments) also depends on the nature of the beach material. As a rule of thumb, coastal engineers space groynes at intervals equal to twice their length on shingle beaches and four times their length on sand. If groynes are spaced too widely, waves arriving at an angle to the coastline drift sand and shingle within each compartment, piling it against the updrift side of

Figure 61 Use of a terminal groyne (arrowed) to trap longshore drift and widen a beach of sand and shingle to protect eroding cliffs on Hengistbury Head, east of Bournemouth in southern England

Figure 62 A harbour breakwater serves as a terminal groyne that has intercepted shingle drifting along the shore to form a high, wide beach for the resort of Fécamp in northern France

the next groyne until it spills over or round, and leaving the rest of the compartment with little or no beach material. Thus each groyne becomes heavily loaded on one side, and may be undermined by wave scour on the other, so that it tilts and may collapse. Similar problems can arise with the height of groynes. If they are too low beach material soon spills over them, and if they are too high updrift accretion and downdrift starvation may load them unequally. Generally the crest of a groyne should stand just above the limit of swash at normal high tides, and decline seaward with the beach profile.

Most groynes have been built as simple structures projecting at right angles to the coastline, separating rectangular compartments. Some of these have been subject to scour by strong seaward currents developing at either end of the beach compartment when storm waves are breaking on the shore, and this can undermine the flanking groynes (Silvester 1974). Other designs have been used to prevent this by adding lateral projections at the seaward ends, as in Y-, T- or L-shaped groynes, or groynes built at an acute angle to the coastline. On coasts where longshore drifting alternates, the retention of a beach may depend on the angle at which a breakwater is built. At New Street, Brighton, on the coast of Port Phillip Bay, Australia, a breakwater was built at an angle of 60° to the coastline in order to prevent cumulative beach accretion at the end of a beach by allowing summer waves to reach the material that had accumulated in the winter season (see Fig. 95, p. 176).

There are merits in an experimental approach to the length, spacing, height and design of groynes built along a beach, inserting temporary structures to determine the probable response of the beach when more permanent groynes are built. Groynes can be lengthened, shortened, raised or lowered in response to beach changes that occur, but it is more difficult to adjust their spacing. An alternative is to carry out experiments in model tanks, using scaled-down waves and beach sediments, before actually building a groyne field, but the limitations of such trials should be borne in mind.

In practice, few groyne fields have been entirely satisfactory, and many have become failures (Brampton and Motyka 1983) (Figs. 63 and 64). Apart from the problems of downdrift erosion, a major difficulty is that beach sediment is lost seaward from the compartments between the groynes (Fig. 65), particularly during storms, when sand and gravel are withdrawn to the nearshore sea floor, and only part of this is washed back during subsequent calmer weather. To avoid such losses, submerged sea walls have been added to some groynes to form enclosed beach compartments (Fig. 66). Where beach sediment is carried away offshore or alongshore from the nearshore zone, a deficit may develop on beaches within a groyne field. Groynes are also damaged during storms, and may quickly become derelict. Failures of breakwaters have occurred at such places as Sines in

Figure 63 Unsuccessful use of groynes on the coast at Rimini, Italy, where they failed to induce sand accretion or to prevent losses of sand from existing beaches

Portugal, Bilbao in Spain, Crescent City, California, and Kahulu, Hawaii. Breaching and collapse usually occur during storms, but the structures have often been undermined and weakened previously by erosion, wave stress, foundation instability, throughflow pressures, or poor durability of the constituent materials. Where the rock used in breakwaters is rapidly weathered or reduced by solution (particularly of limestone), it may settle or topple. Concrete breaks under the pressure of storm waves, especially if it has not been reinforced, and there is progressive fracturing, cracking, and sapping of breakwater and groyne foundations.

There are certainly some coasts where groynes are of no use and should be taken away, leaving the beach to be maintained by unimpeded longshore drifting (p. 182). Alternatively, it may be preferable to maintain beaches by nourishment, rather than build groyne fields.

It is not surprising that, given the many examples of unexpected adverse effects of structures such as breakwaters and groynes on nearby beaches, there is often public opposition to the building of such structures, and also harbours and marinas, on beach-fringed coasts. Proposals to build such structures should be accompanied by a thorough analysis of the processes at work in the coastal sector and the possible consequences for nearby beaches. As such consequences cannot always be exactly predicted, there may be a case for preliminary experimental structures that can be removed if unacceptable effects ensue. There are certainly situations where the

Figure 64 A sea wall was built to form an esplanade at Seaford, in south-east England, behind a depleted shingle beach (see Fig. 45). Wooden groynes, now derelict, and various kinds of concrete groyne were then added in the hope of retaining beach material, but by the mid-1980s the shore was little more than a museum of unsuccessful coastal engineering. The beach was then nourished (see Fig. 89)

removal of a breakwater or harbour would be followed by an improvement on the beaches that have been adversely affected as a result of their construction.

ARTIFICIAL HEADLANDS

The building of breakwaters at intervals along the coast to form artificial promontories or headlands between beach compartments was advocated by Silvester (1976) on the grounds that they would allow the dominant wave pattern to shape a particular beach configuration within each compartment. Essentially a beach would be divided into a succession of smaller bays

Figure 65 Groynes built to retain shingle on the coast at Criel Plage in northern France prevent losses by longshore drifting, but still permit gravel to be withdrawn seaward from the intervening compartments by strong wave action when storms accompany high tides

between the artificial headlands (Fig. 67). Waves arriving at an angle to the coastline would then shape each beach into a 'half-heart' or zeta-curve configuration, which would be relatively stable, with minimal losses alongshore. Each headland groyne would prevent longshore drifting until the intervening beaches had widened sufficiently by accretion to allow sediment to move past. Such structures do not necessarily produce stable beaches.

OFFSHORE BREAKWATERS

Another approach to beach protection has been to build detached structures in the nearshore zone, designed to interrupt and reduce wave action and thus shelter the beach from erosive waves. Offshore (sometimes known as nearshore) breakwaters diminish longshore drifting, and can impede the withdrawal of sediment from a beach by storm waves (Pope 1986).

A well-known example is the breakwater built parallel to the coast and about 600 m offshore to protect Santa Monica pier in 1934. Beach accretion in its lee formed a cuspate spit, and dredging has been necessary to prevent this spit from growing out to join the breakwater.

Figure 66 At Sochi, on the Russian Black Sea coast, groynes failed to retain beach material and so an undersea wall was built to enclose the beach compartments. Beach material is now retained, but on this almost tideless shore the undersea walls are a hazard to boats and swimmers, and the enclosed beach segments retain litter and pollutants

Offshore breakwaters are most effective on tideless shores, such as those of the Mediterranean Sea, where their size and placement do not have to face the problems of a regularly rising and falling sea. They have been in use in Italy for over 70 years, and have become very extensive in the past 25 years. They are generally built parallel to the coastline, between 50 and 100 m long, and 50–200 m offshore, by stacking large boulders, usually of limestone or granite. Much of the Italian seaside resort coast between Venice and Rimini has beaches lined by offshore breakwaters built of limestone blocks brought from the hinterland (Fig. 68). Waves break against them, and are diffracted through the intervening spaces before they reach the beach. This diffraction can lead to the shaping of cusps and shallow zones in the lee of each offshore breakwater, and if there is a sufficient sand supply these may grow into linking tombolos or tombolinos*. Somewhat longer offshore breakwaters (250–350 m) have been constructed to create artificial tombolos on the Baltim Sea Resort beach on the shores of the Nile delta in Egypt (Fanos et al. 1995).

* A tombolo is a cuspate spit that has grown out to attach an island or offshore breakwater to the mainland. A tombolino is a similar feature which is at least partly submerged by the sea at high tide.

Figure 67 Use of headland breakwaters on the shore of a reclaimed area on the south-east coast of Singapore. As expected, wave action formed asymmetrical beaches in the intervening sectors, but these did not stabilise, and continuing erosion required additional shore protection to prevent losses of reclaimed land

Figure 68 A chain of offshore boulder breakwaters on the almost tideless Adriatic coast north of Rimini, Italy, has reduced beach erosion and caused some lee accretion

It is possible to clamber on to the offshore breakwaters, and anglers may fish from them. The boulders often become coated with slippery algal growth or seaweed. There are sometimes objections on the grounds that they disfigure the coastal scenery, interrupt the view of the sea from the beach, and impede boating. The relatively sheltered water in their lee can become more polluted than it would have been with freer water circulation in the absence of such structures, and there is a tendency for beach and nearshore litter and pollutants to accumulate behind each breakwater, in the area used for swimming.

At Ceará, on the coast of Brazil, an offshore breakwater 430 m long was built parallel to an eroding coastline in 1875. It was hoped that this would protect the beach from erosion by waves and still allow longshore drifting of sand from east to west, but the exclusion of wave action also removed the energy for longshore movement of sand behind the breakwater, so that accretion of a cuspate spit began in its lee, and erosion intensified downdrift. Eventually this cuspate spit grew until it reached the breakwater, forming a tombolo. This was then prograded on the eastern (updrift) side, where sand continued to arrive, until it began to spill round the outer side of the breakwater and form a bar trailing from its western end, eventually to rejoin the coastline.

Offshore breakwaters are under construction at Sea Palling, on the north coast of Norfolk, England. Erosion of this coast was severe during the 1953 North Sea storm surge, after which a sea wall was built, but continued lowering of the nearshore area has led to the need for further protection.

Some offshore breakwaters have been built at an angle to the coastline, with the aim of providing shelter from strong wave action without entirely halting longshore drifting in the manner of breakwaters attached to the coast. At Sidmouth, on the south coast of England, angled breakwaters have been built to shelter a nourished beach (Fig. 69) (p. 129).

It may be possible to use mobile, floating breakwaters, which can be anchored in the nearshore zone as a means of preventing beach erosion in phases of strong wave activity, then towed away to allow the beach to be washed and nearshore water cleaned by gentler wave action in fine weather. Various attempts have been made to construct floating breakwaters of rubber tyres, oil drums or timber, intended to reduce wave action and so diminish erosion or promote accretion on the beach.

OTHER NEARSHORE STRUCTURES

Efforts to disrupt and diminish waves approaching a beach, in order to reduce erosion, have included various structures built beneath the sea. South of Niigaata, on the west coast of Japan, where beach erosion had

Figure 69 An offshore breakwater positioned to give some protection to the nourished beach at Sidmouth, Devon, England, from south-westerly waves

been accentuated by local land subsidence due to ground water extraction and the building of extensive sea walls and groynes of concrete tetrapods, a submerged reef was constructed offshore to protect a nourished beach by diminishing the size of waves moving in to the beach. Underwater breakwaters may consist of dumped boulders, submerged ridges of rubber tyres, or forests of artificial (plastic or rubber) seaweed. These provide additional habitats for fish and shellfish, but they can also become a hazard to boating and swimming. At Bridlington, on the east coast of England, artificial plastic seaweed was laid on the sea floor to reduce wave energy and induce sea floor and beach accretion, but the fronds became flattened by wave action, and when they were dislodged by storms they fouled boat propellers. Sea floor sediment accretion was promoted in The Netherlands when similar polypropylene fronds were laid in an area of 9600 m^3 in water 5–15 m deep, but this had little effect on the adjacent beaches.

The chief disadvantage of such schemes is maintenance. After stormy periods it is not uncommon to find beaches littered with rubber tyres or plastic fronds, and the boulders dispersed over the sea floor.

An alternative has been to generate curtains of bubbles rising from a perforated pipe on the sea floor to reduce the size of waves moving in to the shore. This has been used to ease navigation problems in rough weather at Dover harbour in England, but such a system is difficult to maintain on a scale where it can be used for beach protection.

BEACH DEWATERING

It has long been known that a wet sandy beach erodes more quickly than a dry one, partly because wave scour is more effective on saturated sand, whereas dry sand absorbs swash water (Grant 1948) (p. 78). Seepage of groundwater from beaches as the tide ebbs also contributes to beach erosion, increasing seaward sediment movement in the swash zone (Turner 1995). Attempts have been made to reduce the erosion of sandy beaches by inserting a drainage system. Insertion of drains or pipes at right angles to the beach to accelerate seepage has been found to reduce instability on beach profiles in the drained sector (Chappell et al. 1979, Davis et al. 1992), notably on the beach at Dee Why in Sydney, where trials began in 1991 (Hanslow et al. 1993).

An alternative is to insert a perforated pipe and pump out the water, a process known as dewatering (Parks 1989). The effectiveness of this process was discovered accidentally in 1984 at Hirtshals on the north coast of Denmark, when at the suggestion of the Danish Geotechnical Institute salt water was pumped out of a beach to obtain a supply for an aquarium at a nearby Sea Life Centre. It was noticed that the beach prograded by up to 30 m along the drained sector, and gained a metre in height over the pipe zone. This was attributed to the creation of unsaturated sand in the cone of depression down to the buried pipe, into which wave swash percolated, and from which seaward seepage was reduced, lessening backwash scour.

As a sequel in 1985, pumping of water from a 250 m sector of beach at Thorsminde, on the west coast of Denmark, induced 25 m of progradation seaward of the drainage pipe, and when pumping was deliberately halted this accreted beach was soon washed away by the waves (Vesterby 1988). In 1988 a 180 m long pipe was used to pump water from an eroding beach at Sailfish Point, at the southern end of Hutchinson Island, Florida, and within 4 years the drained sector had prograded (Terchunion 1989). In 1993 dewatering of another beach, at Englewood on the Gulf coast of Florida, showed accretion along the drained sector, maintaining the beach profile when storm waves scarped the adjacent beach.

The prospect of raising and widening a beach by dewatering, improving it as a recreational resource and enhancing coast protection, has stimulated much interest. A number of projects are seeking to determine the situations in which dewatering can be a satisfactory and economic means of preventing erosion and augmenting beaches. It is likely to be effective where a high beach water table has contributed to erosion, but it will have little impact on beach erosion due to the other causes discussed in Chapter 3. Dewatering will have a greater effect on beaches of fine to medium sand (grain size 0.1–0.5 mm) than on coarser beach sediments, particularly granules and gravel, which have higher rates of natural percolation. It will also be more effective on beaches of gentle seaward gradient (1:10–1:50) than on steeper or very flat beaches because the cone of depression reaches an optimum width on a gentle slope. The most satisfactory progradation may be achieved on a gentle seaward slope fronting a normally dry upper beach.

Typically, pumping can dry out a zone extending up to 20 m on either side of the buried pipe. There is an obvious risk that a phase of severe erosion by storm waves will expose the pipe, which then becomes useless, obtrusive, unsightly and a hazard to people using the beach, but so far there is no record of this having happened. While the drying of a beach can improve it for recreational use, there is a possibility that such drying will increase losses by wind action. There may be benefits to beaches downdrift when pumping ceases and the induced accretion lobe migrates along the coast.

The relative costs of dewatering and other methods of stabilising a beach have been estimated by MMG Beach Management Systems (UK) Ltd. Expressed as installation costs per metre length of beach, dewatering is estimated at £500/m (plus pumping costs), beach nourishment with sand dredged from offshore about £1000/m, rock armouring about £2500/m and solid concrete walls about £5000/m. Bruun (1989) suggested that although beach dewatering could be successful as a means of reducing erosion under gentle wave conditions, it would be unlikely to prevent erosion by storm waves. Most beach profiles are several times the width that can be dewatered with a single pipe, so that drainage networks may be required. His view was that dewatering could be useful in maintaining short sectors of wider and higher sandy beach in front of hotels and clubs, particularly where costs could be offset by using the extracted water for such features as aquariums, swimming pools and fountains.

Experimental beach dewatering is in progress on Towan Beach, at Newquay in south-west England, a sandy beach in a cliffed bay exposed to swell from the Atlantic Ocean (Fig. 70*). It is largely submerged at most high tides, but at low tide when sea level falls by up to 7 m a wide, gently sloping surface of wet sand is exposed in front of a sea wall. In 1994 it was decided to attempt to build up this beach by a dewatering scheme, and two pipes were inserted, the first near the top to intercept and divert groundwater seepage and runoff from the land, the second across the middle of the beach. The second was a 30 cm diameter pipe 170 m long, set in an excavated trench with a gravel matrix and a geotextile filter to prevent the calcareous sands forming an insulating calcrete. When the water reaches a specified level in the beach a pump begins to operate, and water is withdrawn through a pumping station in an adjacent cove, and discharged by an outfall beside a nearby island

The project, which was undertaken as an experiment, cost £200 000.

---

* I am grateful to Mr. S. Burstow, beach engineer, Restormel District Council, St. Austell, for information on this project and discussion when we visited Newquay to see it in July 1995.

Figure 70 A berm of dry sand over the pipe inserted (along pecked line) to drain (dewater) Towan Beach, Newquay, south-west England

Monitoring of variations in relation to wind, wave and tidal conditions showed that the beach continued to oscillate in level and profile in relation to storm waves and swell, with episodes of scour when strong waves were reflected from the backing sea wall. Nevertheless there were phases in the summer of 1995 when the beach dried out in a zone up to 20 m landward and 10 m seaward of the pipe alignment, forming a slight berm with lower, flatter and wetter sand on either side. There were then plans to extend the drainage pipe laterally, in order to withdraw water from the faint swale landward of this berm. The experiment will be watched with much interest to decide whether dewatering is a viable option on low intertidal sandy beaches on coasts of this type, where tide range is large.

On the Namibian coast a large artificial sand dyke, placed about 300 m seaward of the beach to allow diamond extraction, was kept dry by pumping in order to reduce erosion (Moller et al. 1986, Moller and Swart 1987).

CONCLUSION

Demands for the halting of coastal recession and beach erosion have peppered the world's coastline with an array of artificial structures of various kinds, some of which have been successful, others of little value: many are derelict. On the Lido di Pellestrina, near Venice, attempts were made to retain a wasting beach after a sea wall was constructed, then

Figure 71 After a sea wall and groynes failed to retain the beach at Litorale di Pellestrina, near Venice on the coast of Italy, limestone blocks were dumped on the beach. This is an extreme example of the artificialisation of the coast where engineering structures fail to achieve the aim of preserving a beach

elaborated and armoured with boulders. Groynes were built, then a chain of offshore breakwaters, but as these defences multiplied the original sandy beach washed away, and has been replaced by dumped boulders (Fig. 71). In the words of Dean O'Brien, an eminent American coastal engineer: 'Along the coastlines of the world numerous engineering works in various states of disintegration testify to the futility and wastefulness of disregarding the tremendous destructive forces of the sea'.

O'Brien was discussing the need for specific training for coastal engineers, many of whom are now aware of scenic and heritage values and the importance of nature conservation (Barrett 1993). An alternative to the construction of sea walls, boulder ramparts, breakwaters and groynes is to nourish and maintain beaches as a 'soft engineering' means of absorbing wave energy and protecting the coastline from further erosion (Wiegel 1987). While many coastal engineers favour this option, there is some caution, some doubting the long-term effectiveness of beach nourishment, others querying its economic viability. Where beach nourishment projects fail, as at Morecambe on the north-west coast of England, there is a tendency to revert to 'hard engineering' structures, and protect the coast with massive stone or concrete walls, ramparts and groynes.

The principles and problems of beach nourishment will now be considered.

Chapter Five

Beach nourishment

INTRODUCTION

In the past few decades many beaches depleted by erosion, particularly in the United States, Western Europe and Australia, have been restored by dumping sand or gravel on the shore (Schwartz and Bird 1990). Such sediment, brought from inland, alongshore or offshore sources, has been deposited mechanically or hydraulically to form a beach built higher and wider than the depleted beach (US Army Corps of Engineers 1984). The aim has been to restore and maintain a beach that will protect the coast from storm damage and provide an area for seaside recreation (Dean 1987a).

Beach nourishment (also termed replenishment, restoration, recharge, reconstruction or fill) is artificial in the sense that the sediment has been brought to the shore by engineers. The term beach renourishment is appropriate where a previously nourished beach has been maintained or replenished by further deposition of suitable sediment, and the term artificial beach should really be restricted to situations where there was previously no natural beach. A few artificial beaches have been formed on sectors of coast that had no natural beach, as at Ibiza in the Balearic Islands and Praia da Rocha in Portugal (Psuty and Moreira 1990). Most nourishment projects have taken place on shores where natural beach nourishment (from the sources mentioned in Chapter 1) has diminished or come to an end. The topic of beach nourishment has a substantial and rapidly growing geomorphological and engineering literature, part of which has been reviewed by Davison et al. (1992).

Many beach nourishment projects have been at seaside resorts where beach erosion had become a problem, and the aim was to restore the beach for recreational use. However, in recent years a number of beaches have been inserted primarily as a means of protecting the coastline by absorbing wave energy and so preventing further cliff erosion or damage to coastal property. The nourished beach is used as well as, or instead of, hard structures such as sea walls or boulder ramparts.

The principles and problems of beach nourishment, based on a review of some 90 projects that have been well documented in published or readily available literature from various parts of the world, will now be introduced and discussed. There have been failures, and local conditions must always be taken into account when recommending, planning and carrying out a beach nourishment project. This account is intended to provide background for those concerned with coastal planning and management: the more technical details are available in the Shore Protection Manual produced by the US Army Corps of Engineers (1984) and the Delft Hydraulics Laboratory (1987) manual on beach nourishment.

BEACH NOURISHMENT EXPERIENCE

Several countries have introduced beach nourishment projects in the past few decades, but most well-documented projects are in the United States, the United Kingdom, some European countries and Australia.

United States

In the United States beach nourishment is now widely preferred to coast protection schemes that use hard structures, because so many of these have resulted in subsequent loss of the beach by reflection scour. Although there are reports of beach nourishment at San Pedro in southern California as early as 1919 (Herron 1980), the first well-documented project was in 1922 near the site of the famous amusement park at Coney Island, New York (Hall 1952), when 1.3 million m^3 of sandy material dredged from New York Harbour was placed on a kilometre stretch of island beach. Beach nourishment subsequently became fashionable, and by 1991 over 640 km of beaches on the coastlines of the United States had been nourished at a total cost of about $US 8 billion (Davison et al. 1992).

More than 90 beaches have been replenished on the Atlantic coast of the United States, mostly in response to the severe beach erosion and coastal damage that occurred in major hurricanes during the 1950s, and in subsequent storms, such as the Ash Wednesday hurricane in 1962. Up to 1950 most beach nourishment projects were in the New York and New Jersey area, or in southern California, and were essentially prompted by the local availability of material dredged from harbours and navigation channels. Subsequently beach nourishment has become more extensive on the Atlantic, Gulf and Pacific coasts of the United States, using a variety of sources of sand and gravel, including material obtained from inland excavations, or dredged from estuaries, lagoons or the sea floor. All the major coastal communities on the Atlantic seaboard of the United States coast have nourished beaches, or soon will have (Pilkey 1995).

There have been many studies of beach nourishment projects in the

United States, published notably in *Shore and Beach*, the Journal of the American Shore and Beach Preservation Association, and in the *Journal of Coastal Research.* Maintenance of resort beaches on barrier islands along the Atlantic coast of the United States coasts has posed a number of problems, particularly in Florida where a major coastal tourist industry has developed on barrier beaches along a highly developed coastline (Leadon 1991). There has been a long history of natural opening and closing of tidal inlets, as well as the cutting of entrances to give boat access to backing lagoons, which have been much modified for marinas and navigation canals, causeways and land reclamation.

An example is the city of Miami, which has been built on a sandy barrier island 15 km long and in places only 250 m wide. Miami Beach was originally backed by dunes, but these were levelled to build resort hotels, originally behind a wide beach. Beach erosion became extensive in the 1920s, particularly after southward drifting of sand was interrupted by the cutting of Baker's Haulover Cut in 1925. Miami Beach was further depleted by storm waves during the 1926 hurricane. Individual property owners then sought to protect their frontages by building groynes and sea walls haphazardly, and there was a jumble of such structures by the early 1960s. Further reduction of the supply of southward drifting sand followed the building of more breakwaters to the north, and erosion became severe, Miami Beach losing about 200 000 m^3 per year. The problem was exacerbated by the fact that this is a subsiding coast, with sea level rising about 1.8 mm/year.

By 1975 the beach had become narrow (Fig. 72) and storm waves were washing the sea walls in front of the hotels and condominiums. Over the ensuing decade there was periodic beach nourishment using sand dredged from offshore, over 10 million m^3 of sand being pumped ashore along 17 km of the coast to form a beach over 150 m wide. The aim was to restore the beach and provide a formation that would protect the coastline. The nourished beach also provided a recreational amenity that revitalised Miami's international tourist industry (Finkl 1981).

Most nourished beaches in the United States have been reduced at varying rates by erosion, and some have now been renourished several times. On the Gulf Coast, for example, Dixon and Pilkey (1989) reviewed 110 beach nourishment and renourishment projects on 35 beaches, and found that 23% of them persisted for more than 5 years, 54% for 1–5 years, and 23% for less than a year. In a general review of beach nourishment projects in the United States, Leonard et al. (1990b) found that on the Pacific coast replenished beaches had persisted longer than those on the Gulf and Atlantic coasts, most lasting 1–5 years. This may be due to the ongoing land subsidence along the Gulf and Atlantic coasts (where a rising sea level has been registered on tide gauges), producing less favourable conditions for the survival of nourished beaches.

Figure 72 Miami Beach as it was in 1970, prior to major nourishment

On the Californian coast beach nourishment was facilitated by the use of material obtained from excavations for building sites, road cuttings and tunnels, as well as sediment dredged to maintain harbour entrances or construct marinas (T.D. Clayton 1989). As a sequel to the San Pedro project in 1919, Herron (1980) listed beaches emplaced at Cabrillo in 1927, Santa Monica in 1938, Oceanside and Silver Strand in 1944, Santa Barbara in 1947 and San Diego in 1948. Between 1919 and 1978 there were about 600 coastal nourishment projects in California, using a total of about 108 million m^3 of sediment.

United Kingdom

Beach nourishment in the United Kingdom is part of the coastal protection responsibility of the Ministry of Agriculture, Fisheries and Food (MAFF), working through the National Rivers Authority, and of coastal municipalities: there is no national Coastal Management Authority. These various agencies build coastal defences and structures intended to prevent river and sea flooding, and have re-shaped and nourished selected beaches. In general, coastal erosion and marine flooding are more severe on low-lying parts of the south and east coasts of England, fringed by lengthy beaches, than in Scotland, Wales and Cornwall to the north and west, where harder rocks and higher ground predominate, and shorter beaches occupy bays and coves on generally steep coasts. In contrast to the United States, there are many shingle (gravel) beaches, and shingle beach nourishment has been carried out at a number of sites. The shingle beach west of Rye in Sussex, depleted by longshore drifting to the Rother breakwaters to the east, has been maintained over several decades by extracting shingle from the area of accumulation alongside the breakwaters and taking it back to the depleted sector (p. 179). In the 1970s shingle and sand were brought in from the sea floor to restore beaches at Bournemouth in Dorset, which had diminished after the building of an esplanade cut off the natural supply of sediment from the cliffs. Similar losses of beach material in front of the seaside esplanade at Seaford in Sussex were made good by bringing in gravel dredged from the sea floor and placing it over a seafront armour of large boulders, and the same technique has been used at Sidmouth in Devon, with gravel brought from an inland quarry. Details of these projects are given on (p. 166). There has also been nourishment of sandy beaches on the British coast, as at Great Yarmouth in Norfolk and Portobello in Scotland (Craig-Smith 1973, Beven 1985).

Sea walls and groynes have been widely used on the coasts of England and Wales, and in several places the compartments between the groynes have been nourished with shingle to form a protective beach: an example is seen at Minster, on the Isle of Sheppey in Kent. As has been noted in Chapter 1, several beaches have been incidentally nourished by the arrival of mining tailings washed down rivers, as at Pentewan and Par on the Cornish coast (p. 10), or by spillage of waste from coastal quarries, as at Porthoustock and Porthallow, also in Cornwall (p. 15). Colliery waste has augmented beaches on the Durham coast, quarried gravel has been used to restore the beach at Aberystwyth in Wales, and rubble from a former steelworks has become beach material on the coast at Workington in Cumbria (p. 14).

Europe

Several European countries have introduced beach nourishment as a means of improving recreational beaches and contributing to the protection of their coastlines from erosion. On the west coast of Jutland in Denmark beach nourishment has sought to offset the effects of constructing breakwaters which have intercepted southward drifting of beach sediment, resulting in erosion of beaches downdrift (Thyme 1990, Laustrup 1993).

In The Netherlands much of the coast has beach and dune systems protecting extensive areas of land that are below high tide level. The country is a region of subsiding land, and the response to the gradual rise of sea level over many centuries has been to counter-attack by building and enlarging sea walls (dykes) to enclose and reclaim tidal marshlands and shallow sea areas. Large sea walls now dominate long sectors of the coastline. Where beaches and dunes still exist, the aim of Dutch engineers has been to maintain the coast by nourishing the nearshore zone as well as the beach, and providing sufficient sand to prevent erosion of the dunes behind the beach (Roelvink 1989). Sea defence works have been rebuilt and enlarged in response to recurrent storm surge flooding, and efforts are being made to place beaches on the seaward sides of large walls to give added protection to the structures, provide recreational areas, and soften the artificiality of concrete coastlines. Jelgersma (1975) described an example of this from Neuharlingersiel, in East Friesland.

Other notable beach nourishment projects in Europe have included those on the Polish Baltic coast (Rotnicki 1994), on the islands of Sylt and Nordeney in Germany (Kunz 1990, Kelletat 1992), near Ostend in Belgium (Kerckaert et al. 1986), at Brest (Hallégouet and Guilcher 1990), at Saintes Maries de la Mer and Marseille in France (Rouch and Bellessort 1990), Praia da Rocha in Portugal (Psuty and Moreira 1990), Liso di Jesolo near Venice (Zunica 1990) and Monte Circeo and Marina di Cecina in Italy (Ciprani et al. 1992, Evangelista et al. 1992). Beyond Europe, in the former USSR, Black Sea coast beaches have been restored at Odessa in the Ukraine (Shuisky 1994) and on the coast of Georgia (Zenkovich and Schwartz 1987).

Africa and Asia

In Egypt, sand brought from a desert area near Cairo has been used to nourish beaches at Alexandria (Fanos et al. 1995), and beaches have been nourished on other African coasts at Lagos in Nigeria (p. 178) (Ibe et al. 1991) and Durban in South Africa (p. 178) (Laubscher et al. 1990). In south-east Asia several beaches have been recharged in Malaysia, Singapore and Hong Kong. The Japanese have created beaches on the artificial coast at the head of Tokyo Bay and restored them in seaside resorts such as Niigaata.

Australia

Beach erosion has become widespread in Australia, and state agencies such as the Coast Protection Board in South Australia (Wynne 1984) and the Beach Protection Agency in Queensland (Pattearson and Carter 1989) have introduced beach nourishment schemes. Eroding beaches have been replenished, particularly on the Gold Coast in south-eastern Queensland, on the Adelaide coast in South Australia, and in Port Phillip Bay in Victoria. Brief accounts of these three areas follow.

Gold Coast, Queensland

On the east coast of Australia the predominance of south-easterly swell and storm waves results in northward drifting of beach sand, which has been supplied mainly by rivers, but also partly from cliff erosion and from the

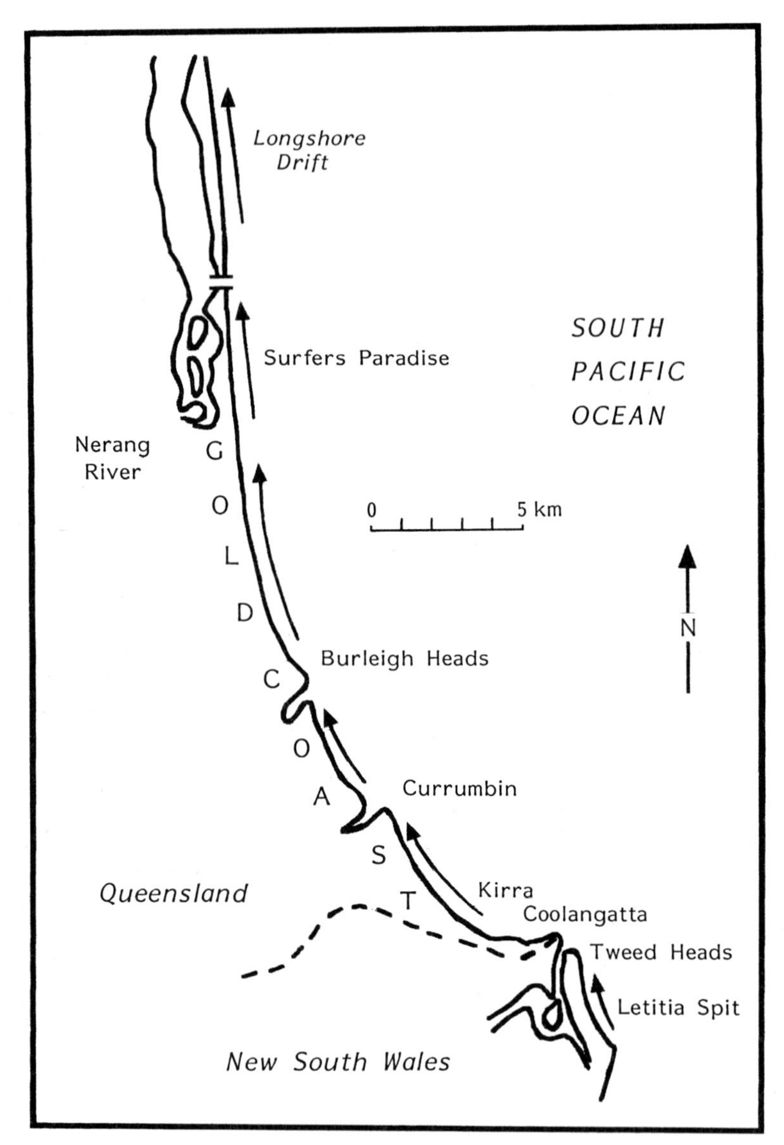

Figure 73 The Gold Coast, south-east Queensland, Australia

sea floor. In northern New South Wales beach sand used to drift past the mouth of the River Tweed and flow intermittently round Point Danger to the Gold Coast, in south-eastern Queensland (Fig. 73), where there has been large-scale tourist resort development, especially since 1950, behind 42 km of sandy beaches. Income from Gold Coast tourism is of the order of $A100 million per year, and as the surf beaches are the main attraction considerable efforts have been made to nourish, maintain and manage them, using sand dredged from estuaries or pumped in from the sea floor (Gale 1979). Nourishment of the Gold Coast beaches is now costing about $A10 million per year, which is seen as an investment necessary to sustain these beaches as an attraction and recreational facility for residents and tourists.

North from Coolangatta the low-lying sandy Gold Coast has been urbanised, with rapid growth of the resort town of Surfers Paradise. Beaches on this part of the coast had been occasionally eroded by storm waves during cyclones, as in 1950, but they had been gradually restored in succeeding calmer weather, when gentler waves washed the sand back on to the shore. When Surfers Paradise beach was depleted by a series of cyclones in 1967 (McGrath 1968), a large boulder wall was built to safeguard the coastal hotels and apartments (see Fig. 50). In retrospect it seems that resort development had been allowed to occur too close to the coastline, and a more generous set-back limit would have allowed the beach to revive naturally during the ensuing few years. Instead, the boulder wall caused scour by reflecting the waves, which lowered the beach and prevented its natural restoration.

In 1970 a report from the Delft Hydraulics Laboratory in The Netherlands recommended major beach nourishment at Surfers Paradise, and in 1974–75 the Gold Coast City Council and the Queensland Beach Protection Agency dredged 1.4 million m^3 of sand from the nearby Nerang estuary, and used it to replenish 3.5 km of beach. This addition of 400 m^3 of sand per metre of coastline formed a new wide beach, with a low dune in front of the sea wall, and these features persisted for several years until they were removed by storms in 1983–84. Renourishment of the Surfers Paradise beach will be necessary at intervals until it can be maintained by the restoration of longshore drifting from the south.

Further details of beach nourishment on the Gold Coast, and on several other beaches nourished in recent years by the Beach Protection Agency of Queensland (Pattearson and Carter 1989), are given under topic headings below.

The Adelaide coast, South Australia

On the Adelaide coast in South Australia there is a 28 km west-facing sandy beach which receives ocean swell and storm waves mainly from the south-west (Fig. 74). These generate northward drifting at the rate of about 30 000 m^3/year, which has fed the spit that has grown northward at Port Adelaide, and produced sectors of accretion at Semaphore and in Largs Bay. There is little sand arriving from the sea floor or from cliff erosion to the south. Suburban development, beginning at Glenelg in the 19th century, has spread along much of this coast, and in the 1970s there was

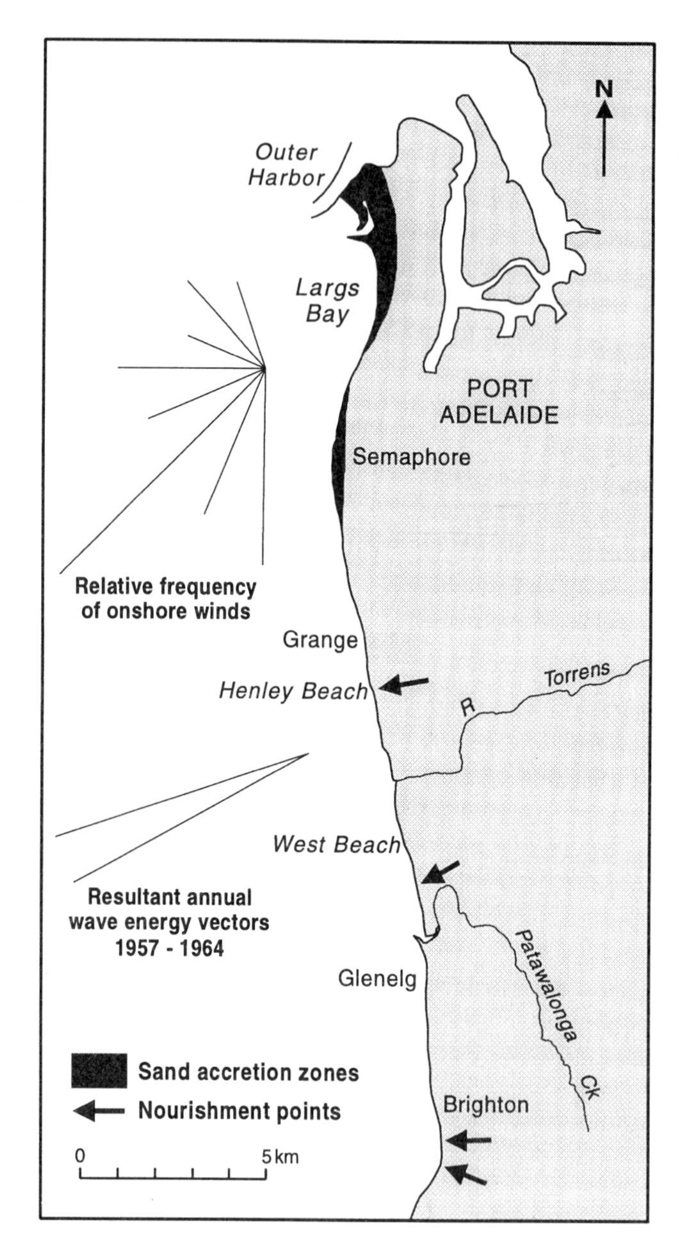

Figure 74 The beach system at Adelaide, South Australia, showing the relative frequency of onshore winds and the range of directions of resultant annual wave energy vectors 1957–1964

Figure 75 Steps ending above the beach at Brighton, on the southern end of the Adelaide coast in South Australia, indicate how it had been lowered by erosion as sand drifted away to the north. A boulder wall was built, and the beach was later nourished

concern at the loss of beaches as the result of a succession of storms. Beaches in the southern sector, between Seacliff and Henley Beach (Fig. 75), were much depleted, and extensive sea walls were constructed to halt coastline recession. The recreational beaches fronting these continued to dwindle, and there were demands that the beaches be maintained. The outcome in 1973 was to initiate nourishment by trucking sand extracted from the accreting areas southward to the eroding beaches (Wynne 1984). This recycling effectively compensated for losses by northward drifting.

Further details of the Adelaide beach nourishment programme are given in later sections.

Port Phillip Bay, Victoria

Port Phillip Bay (Fig. 76) in south-eastern Australia, is a marine embayment with a coastline 256 km long, and since 1975 some 20 of its beaches, averaging 1 km in length, have been nourished (Bird 1990a). The work has been carried out by the Port of Melbourne Authority (formerly the Victorian Division of Ports and Harbors). The results have been favourable, this being a generally good environment for beach nourishment, and most of the beaches have persisted for several years (Bird 1991).

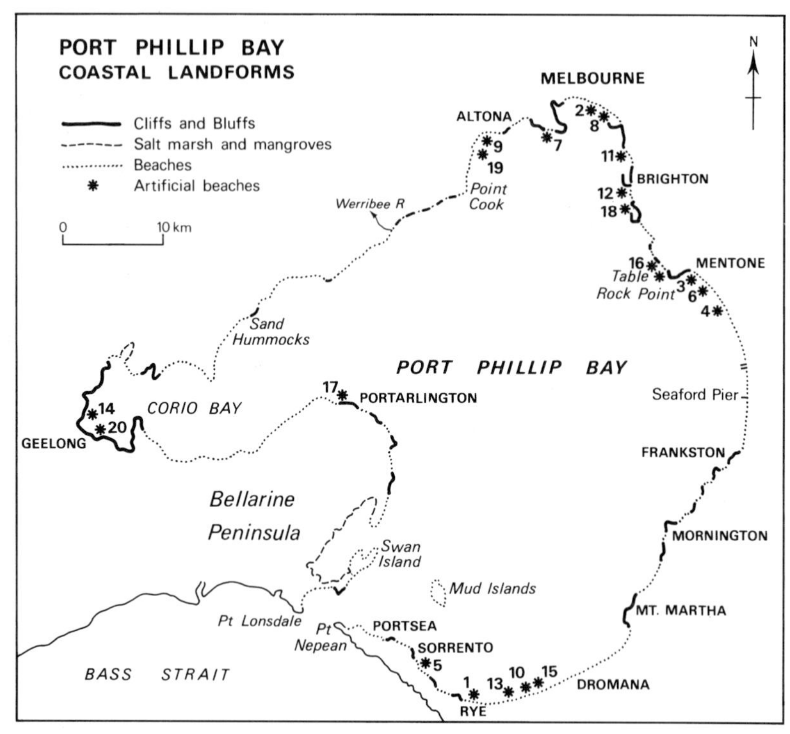

Figure 76 Port Phillip Bay, Australia, showing the location of nourished beaches

Spring tide range in Port Phillip Bay averages a metre, and nearshore tidal currents are generally weak. The ocean swell which dominates most Australian beaches has little effect within this bay, and waves breaking on the bay shores are generated by local winds, especially from the west. Along the eastern coastline waves of up to 1 m are common, and wave heights of up to 3 m occur during storms (Port Phillip Authority 1977).

On the north-east coast of Port Phillip Bay (Fig. 77) the natural beaches are sandy, with some gravel locally. They have been derived largely from the erosion of cliffs and rocky shores cut into soft Tertiary sandstones, because the inflowing rivers, the largest of which is the Yarra, carry only fine sediment (silt and clay) and little sand. Some sand has been washed in from the sea floor during Holocene times, but there is little evidence that this is still continuing, apart from local accessions of shelly material (Bird 1993a). Until recently these beaches were maintained by the supply of sand (and some gravel) eroded from the cliffs.

With the spread of the bayside suburbs of Melbourne there were demands for the stabilisation of the cliffs. Surveys showed that cliff recession at rates of up to 30 cm/year was threatening to undermine the coastal highway (known as Beach Road) at several sites, and between 1936

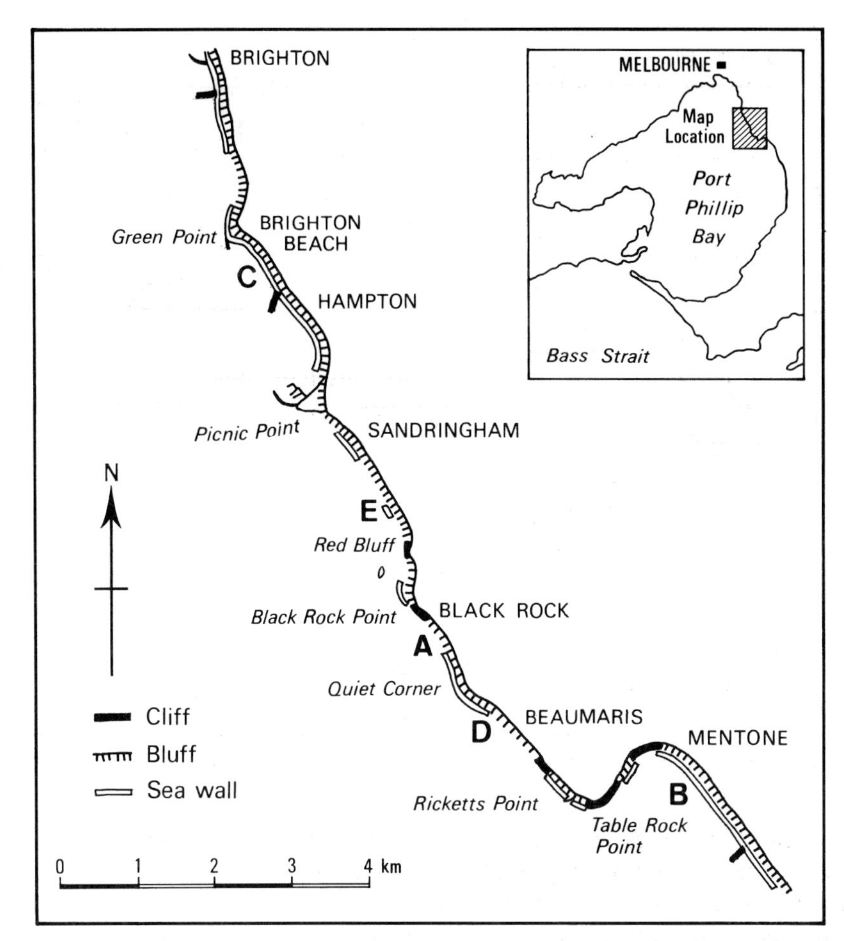

Figure 77 The north-east coast of Port Phillip Bay, Australia, showing the location of nourished beaches mentioned in the text. A, Black Rock Beach; B, Mentone Beach; C, Hampton Beach; D, Quiet Corner Beach; E, Sandringham Beach

and 1946 sea walls were built along the base of several cliffed sectors, which were landscaped to artificial slopes and planted with vegetation. This halted cliff recession, but also curtailed the natural replenishment of sand to the beaches, which then began to diminish. Wave reflection from the sea walls then scoured away much of the remaining sand (see Fig. 49). Experiments with wooden groynes failed to retain the beaches, and the Victorian Division of Ports and Harbors decided in the 1970s to introduce beach nourishment on several sectors, including Mentone, Brighton, Quiet Corner and Sandringham on the north-east coast of Port Phillip Bay.

Details of these projects, and the lessons learned from them, are given in later sections.

PRINCIPLES AND PROBLEMS

Studies of beach nourishment projects can provide a basis for establishing principles and discussing the problems that have arisen. In general, beach nourishment is a feasible method of restoring, protecting or extending a beach while maintaining a natural appearance and conserving coastal processes. It avoids the disadvantages of solid coastal defence systems, such as sea walls and groynes, which can cause reflection scour and downdrift erosion. It may be used independently as a means of protecting the coastline and controlling erosion, or in association with rigid protection structures, especially groynes, which slow down the rate of losses alongshore. The aim of nourishment is not necessarily to restore a beach to some previous condition, but to create a formation that will protect the coastline and persist in the face of wave action.

Manuals on beach nourishment and shore protection have been published by the Delft Hydraulics Laboratory (1987) and the US Army Corps of Engineers (1984), and a general review of the topic has been provided by Hall (1952) and Dean (1983).

The need for preliminary investigations

The failure of a number of beach nourishment projects demonstrates the importance of preliminary research in the planning of beach nourishment (Vallianos 1974, Leadon 1991). An understanding of coastal geomorphology is an essential background for planning beach nourishment, and indeed for any form of artificial modification of a coastline. The diversity of beach systems indicated in the early chapters of this book should be taken into account before deciding how far experience gained from one beach nourishment project can be applied to another, even in the same region. There should be studies of coastal configuration, variations in morphology and sediments, aspect in relation to wind and wave energy regimes, tidal conditions, and particularly the nature and form of the nearshore area. Most nourishment projects are in response to erosion of the natural beach, and it is necessary to know why the natural beach was being eroded (see Table 3), and where the sediment has gone: landward, seaward or alongshore. Project design should take account of the geomorphological context of a beach nourishment proposal, the extent to which it may be influenced by features and processes on adjacent sectors of coastline, and the effects its emplacement may have on those sectors. Headlands, islands, reefs, shore platforms, river mouths, tidal inlets and mobile dunes can all affect processes and dynamics on a coastline, both on the beach to be nourished and along the coast in either direction. There is a need for preliminary research on the movement of sand and gravel in relation to varying wave regimes and the effects of any artificial structures on the shore sector to be treated.

Figure 78 Lorry dumping sand on the shore to nourish Black Rock Beach, Australia

Five examples will illustrate how studies of failing beach nourishment projects can provide a better understanding of coastal systems that can be applied in later projects.

1. At Black Rock, on the coast of Port Phillip Bay, Australia, a sea wall was built to stabilise a cliff in the southern part of the beach compartment (see Fig. 14). The beach then became depleted in front of the sea wall, which was repeatedly damaged by storms, especially in the winter months. In the spring of 1969 the Melbourne Board of Works brought about 5000 m^3 of sand from an inland quarry and tipped it on to the southern part of the beach (Fig. 78) in the hope that it would remain in place to protect the sea wall. The effect was to increase the total volume of sand on this beach by about 10%, and in the summer months the nourished beach sector remained wide. During the autumn, however, the dumped sand was carried away by northward drifting, leaving the sea wall again exposed to the impacts of winter storms. The project had failed to take account of the pattern of longshore drifting, which alternates seasonally on this coast, and showed that much more sand would have to be dumped to ensure that a protective beach would persist on the southern sector, in front of the sea wall, through the winter months.

Figure 79 Nourished beach at Cliftwood, New Jersey

2. At Wrightsville, North Carolina, a beach was nourished in such a way that waves coming in at an angle to the shore caused rapid loss of the emplaced sand by longshore drifting into a neighbouring inlet. When a terminal breakwater was built to reduce losses into the inlet, the intercepted sand prograded the beach to a new alignment, more closely related to the pattern of incoming waves. Longshore drifting then diminished, and the beach became relatively stable. The project showed that it was necessary to nourish a beach to a correct alignment in relation to incident wave action. A beach built parallel to the prevailing waves should lose little sand by longshore drifting, and one built at right angles to the resultant of wave orthogonals (i.e. with wave action from one side balancing that from the other) should show no nett longshore drifting. Such beaches can still lose sediment seaward in response to destructive waves.

3. At Cliftwood, New Jersey, beach material dumped on a convex beach salient in 1985 was quickly carried away alongshore by divergent drifting. This is a sinuous north-facing coast (Fig. 79), and when Jackson and Nordstrom (1994) mapped patterns of erosion along the nourished beach, they found that the redistribution of the emplaced beach material by longshore drifting was related to changes in shoreline orientation. Beaches in bays or re-entrants proved to be more stable and persistent than those on straight or convex sectors of the coastline,

the convex salients quickly losing sediment alongshore. It is difficult to maintain a nourished beach on such salients because the protruding coast is exposed to waves from several directions, and is likely to lose sediment quickly. On the other hand, nourishment on a salient could be a way of nourishing the whole of a beach compartment, allowing longshore drifting to carry sand into the intervening bays.

4. On the island of Langkawi, in Malaysia, coastal hotels purchased sand from inland quarries to improve their beaches individually, but this produced an irregular pattern of lobate protrusions, which subsequently migrated away northward along the coast. There is little point in nourishing one or more small sectors for local protection within a beach compartment. It is preferable first to study the beach and nearshore system within the compartment, and decide where locally deposited sediment would go, in order to devise an integrated beach nourishment project rather than this piecemeal operation.

5. At Morib, on the southern shores of the Klang delta in Malaysia, longshore drifting carried sand from the beach away northward to form a spit curving in behind a mangrove swamp. A beach nourishment project, using 12 000 lorry loads of sand brought back from the spit, failed because a sea wall had been built behind the depleted beach, and this increased reflection scour when waves broke against it at high tide, sweeping away the deposited sand. This demonstrated the need to assess the effects of artificial structures that have been added to the coastline, in order to decide how these may influence the nourishment of a beach and its subsequent performance, before introducing beach nourishment.

Geomorphological and engineering expertise are thus necessary to decide how beach nourishment should be carried out. Studies are required to determine the nature and volume of sediment to be injected, and how the renourished beach is likely to behave in ensuing years. Plans to renourish a beach should benefit from experience gained from the previous nourishment project, particularly concerning the patterns and rates of loss of the earlier beach fill and the changes that occurred in grain size characteristics. This will be easier if there has been adequate mapping and monitoring of the changes that took place after the previous beach nourishment (p. 192). The aim is then to select methods of nourishment and types of sediment that will establish a more persistent beach.

An example of a renourishment project guided by mapping and monitoring comes from Rockaway Beach, New York. In 1975, after half a century of periodic nourishment with sand dredged from New York harbour failed to offset continuing losses, the beach was investigated by making numerous surveys of cross-profiles from the backshore to the nearshore zone, and analysing the associated grain size distributions. On this basis it was decided that the beach would be more durable if it were initially shaped to form a profile 30 m wide and 3 m high, with an upper slope of 1:20 declining to a lower slope of 1:30 (Nersesian 1977). Sediment was obtained from the sea floor because it was coarser and better sorted than the harbour sediment that had previously been used and had rapidly dispersed (Hobson 1977). Sand was dredged offshore and taken in by barges to the beach, where 4.3 million m^3 were dumped on the shore to form the new beach profile. The project demonstrated that a more persistent beach could be produced by designing suitable cross profiles and using material of more appropriate grain size.

In the United States, coastal engineers have given much attention to theoretical modelling of beach systems as a basis for designing beach nourishment and predicting subsequent performance. Examples of such symbolic analysis have been documented by Perlin and Dean (1985), Brampton and Motyka (1987), Hanson (1987), Larson et al. (1990), Larson and Kraus (1991) and Hales et al. (1991). By contrast, Dutch engineers have relied on a more empirical approach, based on monitored histories of coastline changes (Delft Hydraulics Laboratory 1987, Van de Graaff et al. 1991). The Netherlands have good background data, which most other countries lack. Beach replenishment projects are there planned on the basis of prior monitoring, calculating the sediment losses that have occurred, and assuming that these will continue. Dutch engineers often overfill a beach by up to 20% to take account of beach profile adjustment and losses following nourishment. Similar techniques were used to assess rates of longshore drifting on the Gold Coast in eastern Australia, correlated with the quantity of northward-drifting sand intercepted by the breakwaters at Tweed Heads, at 500 000 m^3/year.

It has been argued that the complexity and variability of coastal systems are such that the Dutch experimental and ongoing approach to beach nourishment, based on accumulated experience, may be more realistic than theoretical modelling (Pilkey and Clayton 1989), and this approach has been used at Virginia Beach and Wrightsville Beach on the Atlantic coast of the United States.

Monitoring of changes on nourished beaches as a basis for further projects should include calculations of the budget of beach gains and losses over specified periods. Existing monitoring is often inadequate for evaluation of nourished beach performance (p. 192). Lack of documentation of beach changes, both before and after nourishment, has resulted in unreliable estimates being made, using short-term measurements, particularly during the phase of rapid change which often occurs immediately after beach emplacement. It is hazardous to rely on unscientific data such as media reports. Long-term monitoring is necessary to avoid misleading measurements based on seasonal or short-term variations, and to provide for the more effective design of subsequent beach nourishment projects.

Sources of material for beach nourishment

Almost any kind of durable material of suitable grain size can be used for beach nourishment, providing it does not contain pollutants or hazardous items such as broken glass or jagged metal. Sediment is obtained from sites known to engineers as 'borrow areas', a curious term which is only really appropriate in situations where the sand and gravel are expected to drift back to where they came from. It would be better to refer to them as 'source areas'. Ideally the material should have similar grain size characteristics to the natural beach sediment (James 1974); if not, allowance must be made for the rapid removal of the finer constituents as waves re-work the emplaced beach.

Sources of material for beach nourishment should be sought as close as possible to the sector to be replenished, in order to minimise transportation costs. A beach is more likely to be nourished if a source of suitable material exists within an economic distance. At Bournemouth the ideal beach fill was available in the form of china clay residue from Hensbarrow Downs in Cornwall, but this was over 200 km away, too far to be economically worthwhile, and it was necessary to use sand and gravel dredged from the nearby sea floor (Willmington 1983).

Sediment from land quarries

Sand and gravel have been obtained from quarries and trucked down to the coast to nourish beaches in Monterey Bay and Ediz Hook on the Pacific coast of the United States, Michigan City on the Great Lakes, Sidmouth on the south coast of England and Redcliffe in Queensland. Black Rock beach, on the shores of Port Phillip Bay, was partly nourished in this way (p. 147). Erosion on Bramston Beach, Queensland, was offset by bringing in 6000 m^3 of coarse sand quarried from old beach ridges a short distance inland and delivered to the shore by lorries. Acquisition of land-based sources of material for beach nourishment is not hampered by sea conditions, but there may be problems with lorry traffic. However, the cost of using such material for beach nourishment may be too high where there are competing demands for sand and gravel extracted from quarries for other purposes such as road making or concrete aggregate.

Sediment from other beaches

Sand or gravel can be brought alongshore from other beaches, particularly on prograded sectors, to nourish a depleted beach. At Aberystwyth the

Figure 80 Dumping of lorry-loads of sand brought from the prograded sector north of the harbour breakwaters at Hvidesande, Denmark, to restore the eroded beach to the south

resort beach was augmented in 1963 by shingle brought from a beach to the north (So 1974). At Hvidesande, in Denmark, southward drifting sand accumulated alongside breakwaters built in 1910 to stabilise the entrance to a coastal lagoon known as Ringkøbing Fjord and ensure ship access to the port of Ringkøbing on the shores of that lagoon. Sand extracted from the widened beach north of the breakwaters was trucked southward to nourish the downdrift beach, which had been cut back up to 45 m as the result of the drift interception (Fig. 80) (Møller 1990). This is an example of bypassing (p. 175): sediment moved from an accreted area downdrift back to a depleted area updrift is termed recycling (p. 179).

Sediment from harbours

A major source of sediment for beach nourishment has been material dredged from harbours and port approaches. One of the earliest examples of beach nourishment on the Atlantic coast of the United States was in 1925 at Rockaway Beach, on the sandy barrier island coast south of New York, where the insertion of numerous groynes on 16 km of coastline had failed to halt erosion of the sandy beach. Sand was then dredged from New York harbour and dumped on the Rockaway shore. It gradually washed away, and there was periodic renourishment with sand from the bottom of the harbour until 1975, when the beach was renourished with sand brought in from the sea floor to the south.

Sand dredged to maintain a navigable boat channel at Barnegat Inlet, New Jersey, was used in 1979 to nourish the eroded ocean beach on nearby Long Beach Island (Psuty 1984), and 2 million m^3 of sand obtained from a similar boat channel was used to restore the beach on Sandy Hook, New Jersey, where the spit had been breached by erosion in 1982.

Sand dredged from the harbour at Rio de Janeiro in the 1940s was used to nourish Copacabana Beach, which was meagre until it was thus enlarged to provide a wide shore recreational area above normal high tide level (Vera Cruz 1972, Leatherman 1986). At Robe, in South Australia, sand dredged to maintain the port approach has been dumped to restore an adjacent beach.

Several beach nourishment projects in New Zealand have used sediment available as a by-product of port dredging or marina projects (Healy et al. 1990). At Omaha Spit, north of Auckland, New Zealand, an eroded beach was restored using sand dredged from shoals in Whangatau Harbour, to the rear of a spit. Dredging at the ports of Napier, New Plymouth and Tauranga Harbour produced large quantities of sand and gravel, which have been used to nourish beaches downdrift.

Sediment from lagoons

On barrier island coasts the backing lagoons may provide a suitable source of sediment for beach nourishment. Some 443 400 m^3 of relatively coarse sand dredged from a lagoon floor behind a coastal barrier was used to nourish an eroded beach north-east of Atlantic City, New Jersey, in 1963. Sediment was dredged from backing lagoons to nourish beaches on the outer shores of Hel Spit in Poland, while at Beachport, South Australia, sand dredged from a lagoon entrance has been used to nourish the town beach.

Erosion of the beach, at Lido di Jesolo, Italy (p. 109), on one of the barrier islands fronting the Venice Lagoon, was countered by building sea walls, groynes and offshore breakwaters. These failed to retain the beach, which was then replenished by dumping sand dredged from the floor of the lagoon (Zunica 1971, 1990). Sand and gravel dredged from coral lagoons has been used to nourish eroding resort beaches on cays, as at Green Island, off Cairns in north-eastern Australia.

Sediment from rivers

Sediment dredged from Caucasian rivers has been used to nourish beaches on the Georgian Black Sea coast (p. 170), and sand obtained from the estuary of the Nerang River has been used on Gold Coast beaches in

Queensland (p. 178). On the shores of the Barron delta north of Cairns, Australia, the beaches were naturally supplied with sand washed out from distributaries of the Barron River, and eroded when a change in the position of a river mouth brought this supply to an end. Holloways Beach, for example, received Barron River sand from a distributary outlet until a flood in 1938 switched the main outflow northward, through Richter Creek. Deprived of a sand supply, Holloways Beach began to erode, the sand drifting away northwards. In 1992 a 600 m sector of this beach was nourished with 83 000 m^3 of sand dredged from the lower reaches of Richter Creek and delivered to Holloways Beach by lorries, essentially an artificial revival of the fluvial sand supply that originally built this beach. Maintenance of such deltaic beaches could also be achieved semi-naturally by relocating mouths of river distributaries so that they feed sand to the updrift end of a resort beach.

Sediment from tidal inlets

Sediment taken from nearby tidal inlets has been added to beaches at Bournemouth in England and Caloundra in Queensland. When sand is sought for beach nourishment projects on barrier island coasts there are often shoal deposits at tidal inlets, formed seaward by ebb tides and lagoonward by flood tides. It is necessary to consider the possible impacts of extracting sediment from such shoals, such as the effects on wave and current regimes in the inlet, and the possibility that changing wave patterns will have adverse effects on the adjacent coastline. At Captiva Island, Florida, resort beaches were nourished with sand dredged from ebb-tide shoals off the nearby inlet at Red Fish Pass. Its extraction proved to be beneficial because it modified wave refraction patterns in such a way as to impede the drifting of sand back from the nourished beach into the inlet. Sand was also available from flood-tide shoals just inside the inlet, but this was not used because of the risk that deepening this area would induce sediment inflow and result in erosion alongside the inlet (Walton and Dean 1976).

Sediment from the sea floor

In recent years, material dredged from the sea floor has been widely used in beach nourishment in such places as Seaford in England, Ostend in Belgium, Port Dickson in Malaysia, and Miami Beach in Florida. Nearshore shoals have provided sandy material for beaches on Nordeney in Germany, on the Adelaide coast in South Australia, and Brest in France. There are sometimes alternative demands for sediment extracted from the sea floor (e.g. for aggregate or building purposes), but this is less likely

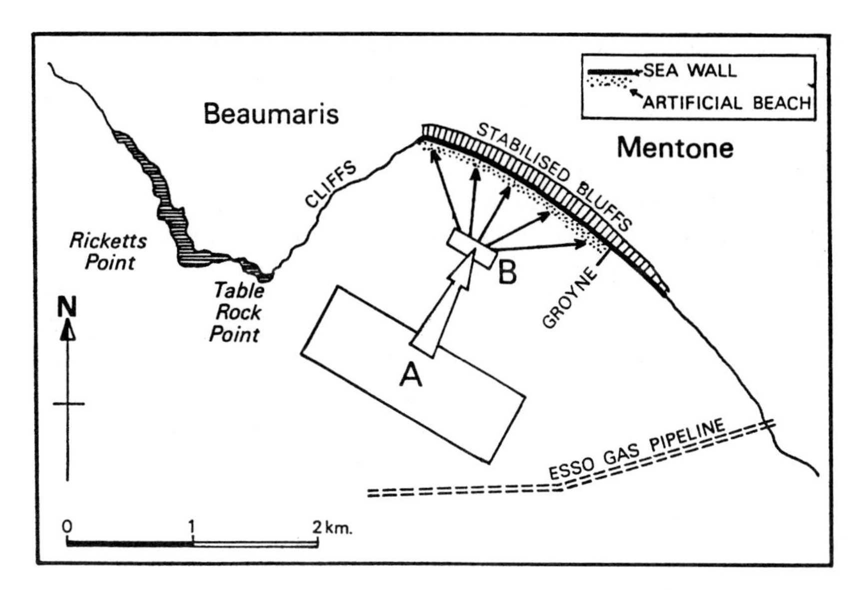

Figure 81 The coast at Mentone, Port Phillip Bay, Australia, showing the area from which sea floor sand was dredged (A), and the zone where it was placed (B) to be pumped on to the shore and nourish the beach, as shown in Fig. 85

where the material contains clay or shelly debris, which can be used for beach fill.

Sand dredged from the sea floor was used in the restoration of Mentone Beach, Port Phillip Bay, Australia. Sea floor sand immediately offshore was too fine and silty to be used for this purpose, but between 1.4 and 2.0 km seaward, where the water was 7–10 m deep, there was a deposit of coarser sand up to 1 m thick, containing marine shells (Guerin 1984). Sand dredged from a rectangular zone 1800 m long and 600 m wide (Fig. 81) was then dumped inshore, to be pumped on to the beach.

Substantial losses of sea floor material to be used for beach nourishment can occur during its extraction, transport and delivery to the shore. The Rockaway Beach project in 1975–77 lost 10% of the volume of sediment originally excavated from the sea floor before it reached the beach. Some was lost during the dredging process, some as the barges were filled, and some as they were emptied on to the shore. Such losses, mainly of fine-grained material, modify the grain size distribution of the sediment. At Rockaway Beach the losses of fine-grained sediment had been allowed for, as relatively coarse material was required for beach renourishment (p. 149).

On the German island of Sylt 1 million m^3 of sand was extracted from the sea floor, but more than 20% of it (also mainly the fine fraction) had been lost before the balance of 770 000 m^3 actually reached the shore (Dette 1977). At New River, North Carolina, a loss of 16% of the original nourishment material left coarser and better sorted sediment, which helped to improve the performance of the restored beach (Hobson 1977).

Sediment from distant sources

There are a few examples of beach sediment being shipped in from a distant source. Oolitic sand dredged from the Bahama Banks was delivered to beaches on the Atlantic coast of Florida (Adams 1979), but unfortunately the cost is high, and it is cheaper to use algal and foraminiferal sand dredged from Florida coastal waters, taking care to avoid damaging coral reefs (Olsen and Bodge 1991).

Coarse sand and gravel dredged from coralline shoals, reefs and cays and ferried in to the coast has been used for beach nourishment on a number of resorts on Caribbean islands. At the seaside resort of Varadero, on the Hicacos Peninsula north-east of Havana, Cuba, the beach suffered severe storm damage in the winter of 1986–87, when waves washed along the hotel fronts. In 1990, 81 000 m^3 of coarse sand was obtained from Cayo Mono, an uninhabited coral cay 25 km offshore, brought in by barge, pumped onshore, then shaped by bulldozers to form a recreational beach 1.4 km long (Schwartz et al. 1991).

Sand was brought from a distance to nourish the beach at Waikiki, Honolulu. Most of the south coast of Oahu, from Pearl Harbour past Honolulu to Koko Head, was originally fringed by a coral reef, much of which has been concealed by land reclamation. The natural beach at Honolulu was originally rather meagre on a rocky, reef-fringed shore, until it was nourished in several stages beginning in the 1920s. There are conflicting reports on where the sand came from. Possible nearby sources include other sandy beaches on Oahu and nearby islands, and sand dredged from Pearl Harbour, but apparently some sand was shipped in from Manhattan Beach, in California, and there are anecdotes of importation from various other sources, including Australian beaches (Campbell and Moberley 1984).

Sediment from mining and quarrying

At Rapid Bay, in South Australia, 1.5 million tonnes of quarry waste (limestone gravel) were dumped between 1940 and 1982 on a previously rocky shore, forming a beach up to 230 m wide and rising 3 m above high tide level. Dumping then ceased, and the beach was cut back rapidly until its seaward edge attained a gentle concave profile (Bourman 1990).

An example of beach nourishment prompted by the need to dispose of mining waste has been reported from Chañaral Bay in southern Chile (Paskoff and Petiot 1990). Between 1938 and 1975 tailings from a copper mine were dumped on the shores of Chañaral Bay, on the Atacama Desert

Figure 82 Beach augmented by the dumping of colliery waste at Ness Point, near Seaham in Durham, north-east England. Dumping ceased when the nearby colliery closed in 1993, and the beach is gradually being re-shaped by wave action, the waste material being sorted into sand and gravels and cleaned by the sea. A low serrated clifflet (arrowed) marks scarping at the limit of wave re-working

coast, at the rate of more than 4 million m^3/year. They included silts and clays, which were swept offshore, and a sandy fraction which prograded the beach by an average of 900 m over this period. After 1975 the dumping was transferred to Caleta Agua Hediona, the next bay to the north, which had a rocky shore with stacks in front of cliffs. By 1985 this had acquired an artificial beach more than 5 km long and up to 300 m wide.

As has been noted (p. 151), a major potential source of material suitable for beach nourishment in south-west England exists in the tip-heaps of quartz and felspar sand and gravel in the china clay quarrying region of Hensbarrow Downs, near St. Austell in Cornwall. There is a similar, smaller area in south-western Dartmoor, north of Plymouth. Transportation costs have so far prevented much use of this resource, although a small beach at Torpoint, near Plymouth, has been replenished with it.

Colliery waste has been dumped on the shores of County Durham, England, for more than a century, until the closure of the coal mines in recent years (Hydraulics Research Station 1970, Nunny 1978). Berms of dumped waste have been reworked and drifted alongshore for several kilometres, augmenting the natural beaches of limestone gravel and locally derived sand (Fig. 82). The additional coal and shale have been sorted into gravel and sand, and in some places coal fragments are being collected from the shore, mainly for domestic use. The wide, convex berms stand in front of limestone cliffs, which have become weathered and partly vegetated since wave action no longer reached their base. Now that dumping has halted, the convex berms are being cut back by wave action, forming a low clifflet or beach scarp (p. 39), fronted by a developing concave profile, and in due course this will extend back to the cliff base, which will again be attacked by waves, rejuvenating and steepening the limestone cliffs.

Waste from a steelworks has been used to nourish beaches at Port Lincoln and Port Augusta, in South Australia. At Odessa, on the Ukrainian Black Sea coast, the material used to replenish eroded beaches included some 3 million m^3 of sediment excavated for the building of Port Yuzhniy, in a nearby coastal lagoon, 30% of which was limestone and the rest sand and clay (Shuisky and Schwartz 1979). This was mixed with urban rubble waste from roads and buildings, and dumped to form an artificial terrace in front of a cliff that had been subject to recurrent landslides to the west of the port. As expected, beach material was generated as the rubble mixture was reworked by wave action, and sand and gravel washed out of the terrace drifted westward to form a new beach in front of previously eroding cliffs in Odessa Bay. The rubble waste thus proved to be a satisfactory source of beach material, transported, sorted and shaped by wave action (Shuisky 1994).

At Oriental Bay, on the north-east shore of Wellington Harbour, New Zealand, a beach was artificially nourished with sand in the 1920s, but this had disappeared by 1935. In 1944–45 about 11 500 m^3 of sand and gravel from ship's ballast was added to the beach to improve the recreational facility, but this has also been reduced, and the present beach consists largely of residual pebbles.

Grain size and nourished beach stability

It has been found that beach nourishment is more effective when an appropriate grain-sized sediment is used (Newman 1976). Beach fill should have similar grain size characteristics to the natural beach, for excessively fine sediment is soon lost offshore or alongshore, and excessively coarse sediment may form too steep a beach, promoting nearshore reflection scour (Krumbein 1957). In general, coarser material is likely to persist longer on a beach, and finer sediment to be lost more quickly (Dean 1983). Monitoring of nourished beaches on the shores of Lake Michigan, in the United States, has shown that those formed of gravel (pebbles or cobbles) last about five times as long as those of fine sand, which disappears quickly offshore (Roellig 1989).

Rapid losses have occurred on beaches nourished with unsuitable material. In 1982 the 12 km beach fronting the seaside resort at Ocean

City, New Jersey, was renourished using poorly sorted shelly sand dredged from a nearby tidal delta, but this was quickly washed away by storm waves (Pilkey and Clayton 1987). An example of the successful use of coarse sediment after finer sediment had been washed away occurred on the Presque Isle Peninsula on the Pennsylvania shore of Lake Erie. A beach that had been losing sand was nourished in 1960–61 by dumping 525 000 m^3 of sand, but this was soon depleted because the fill material had a higher proportion of fine sand than the preceding natural beach. When the beach was renourished in 1965 with 12 700 m^3 of coarser sand it became more stable (Berg and Duane 1968).

Demands for the restoration of Hampton Beach, Victoria, Australia, resulted in 1975 in the pumping of sand from the floor of nearby Sandringham Harbour through a pipeline on to the shore. The nourished beach was washed away by storm waves within a few days, because the sand taken from the harbour floor was much finer than that on the natural beach. This was because the sand that had drifted into the harbour had previously been sorted by wave action, which had left coarse sand on the beach and withdrawn the finer fraction to the sea floor, whence it was transported by nearshore currents into Sandringham Harbour (Bird 1993a).

Several other beach nourishment projects have shown that if the sand used in beach nourishment is too fine it will quickly wash away. At Wrightsville, North Carolina, sand dredged from a nearby estuary was used to nourish the beach five times between 1939 and 1970. In each case the sand was removed during storms and deposited offshore (Pearson and Riggs 1981), and it was realised that the sand taken from the estuary was too fine for long-term retention. Renourishment with coarser material in 1972 was somewhat more successful, although there were still losses to seaward, especially during storms.

Other examples of beach nourishment failing because the sediment deposited was too fine in texture have occurred at Rosslare in Ireland and on the German North Sea coast island of Nordeney. On the shores of Rosslare Bay, County Wexford, in south-east Ireland, beaches and dunes had been eroding at the rate of up to 1.3 m/year over the past century, probably as the result of the building of the Rosslare Harbour breakwater in 1882, which traps longshore drifting from the south. Sea walls, boulder ramparts and wooden groynes were constructed, but failed to stop erosion, and in 1982 beach nourishment was attempted, using 500 000 m^3 of sand dredged from an offshore shoal and dumped at a point on the beach. This proved to be too fine to be retained on the shore, and was soon washed back into the sea (Carter 1988).

On the island of Nordeney the authorities nourished the beach in 1951 using sand dredged from nearshore shoals. The dredged sediment included a higher proportion of fine sand than was present on the natural beach, and this was soon extracted by wave action and washed back out to the nearshore zone (Eitner and Ragutzki 1994). Later renourishment projects have taken account of the fact that this sorting would occur, and that only the coarser fraction would persist on the beach, the finer material moving out to maintain nearshore bar formations (Kunz 1990).

In north Queensland, Australia, the coarse sand quarried from beach ridges behind the coast and used to nourish an eroding beach at Bramston, near Innisfail, was more likely to be of suitable grain size for beach nourishment than sediment which was not originally a marine deposit. This is because it is less likely to be subject to rapid sorting by wave agitation when placed on the shore, with substantial losses of the finer fraction, than sediment which has already been wave-worked.

Selection of grain size in material used for beach nourishment should also take account of wave energy conditions. This was illustrated on the often stormy shores of the Gulf of Georgia near Vancouver, Canada. Sand was introduced in an attempt to form a beach that would stop the erosion of cliffs cut in unconsolidated glacial deposits, but it was soon dispersed by storm waves. A second project, using cobbles, was more successful in establishing a beach to protect the cliffs (Downie and Saaitink 1983). In general, coarser sediment is needed to stabilise a beach in a high wave energy environment, but fine sand may persist on a low wave energy shore. At Nunns Beach, adjacent to Portland Harbour, Victoria, Australia, fine sand that had accumulated south of the harbour breakwater was pumped under the entrance and dumped on the beach in 1990. In this sheltered environment the nourished beach of fine sand has diminished slowly.

The durability of a beach may depend on the shape of the sand or pebbles used, angular material being less readily transported and lost than well-rounded material. In the 1960s an experiment was conducted in St. Lucie County, Florida, when 1000 tons of imported oolitic sand from the Bahama Banks and a similar quantity of local beach sand were placed in rows across the shore. These were redistributed by wave action during high tides, and it was found that the more angular oolitic sand was less mobile than the well-rounded native sand, and therefore more likely to persist on a nourished beach (Cunningham 1966). It has been suggested that waste from coal mines, mostly rocky debris of varied size and irregular shape, unlikely to be quickly rolled away by wave action, could be used to stabilise eroding beaches elsewhere, as in front of the rapidly receding cliffs of Holderness on the Yorkshire coast.

It is also necessary to take account of rates of weathering of beach material, which can reduce its grain size. On Delray Beach in Florida it was found that organic sand dredged from the nearshore zone was brittle, and when placed on the beach it disintegrated rapidly under wave agitation to finer material that was quickly dispersed. The specific gravity of introduced sediment is another relevant factor. In Tauranga Harbour, New

Figure 83 Beach improvement at Whitburn Bay, Durham, north-east England. Sand has been placed over cobble gravels to provide a recreational beach for summer visitors

Zealand, a nourished beach contained pumice, but this very light material was soon washed away, indicating that wave action selectively removes lighter as well as finer material (De Lange and Healy 1990).

Where gravel dredged from the sea floor contains sand and silt its use in beach nourishment can result in the formation of an excessively compact and impermeable beach capping, which may develop a firm cliff near the high tide line. Examples of such scarping have been documented from nourished beaches at Whitstable and Hayling Island in south-east England, where it is regarded as a hazard to beach users, and a cause of reflective scour by waves on the lower beach during storms (McFarland et al. 1994).

Suitability of material available for beach fill was a problem at Aberystwyth, where in 1963, 1530 m^3 of waste from a quarry on nearby Constitution Hill was dumped on the shore to improve the beach at Victoria Terrace. It contained a high proportion of laminated and soft shale, which disintegrated and was quickly dispersed (So 1974).

At seaside resorts the losses of sandy material during winter storms leave a gravelly beach, and it may be necessary to bring back sand to restore the beach for recreational use. At Whitburn Bay, on the Durham coast in north-east England, the resort beach is improved each spring by dumping sand over the gravelly shore (Fig. 83).

Methods of beach nourishment

Methods of nourishing a beach vary according to the configuration of the coast and the processes at work on it. There is direct placement of fill material on a coastal sector, particularly where longshore drifting is weak, or can be controlled by the insertion of breakwaters of some kind. There can be nourishment at points or sectors from which it is expected that longshore drifting will carry the sediment to where it is required. There is bypassing, where sediment is conveyed from a sector that has prograded alongside an obstacle such as a breakwater, a river mouth or a tidal inlet, along the coast to replenish a beach depleted downdrift. There is recycling, whereby beach losses due to longshore drifting are made good by bringing back the sediment. Some nourishment projects include the use of groynes or breakwaters to retain the deposited sediment. There is nearshore nourishment based on the expectation of shoreward drifting, and backpassing, which is analogous to longshore recycling in that it brings back sediment lost seaward from a beach. Offshore breakwaters have been used to nourish a beach by inducing accretion in their lee, and some beaches have been emplaced or reshaped by bulldozing. Each of these techniques will be considered and exemplified in the following sections.

Direct placement

The nourishment of a beach by dumping sediment brought from an inland quarry or from alongshore or offshore sources on a sector of coastline is well illustrated by the beach nourishment programme at Bournemouth, on the south coast of England. The beach here was naturally supplied with sand and shingle eroded from backing cliffs in soft Tertiary sandstone, capped by Pleistocene gravels, which fringed the shores of Bournemouth Bay. Between 1907 and 1975 these cliffs were progressively stabilised by building and extending a sea wall and basal promenade. These cut off the sediment supply, and the beach soon began to diminish. The former cliffs have been artificially landscaped to produce coastal slopes and planted with vegetation (Lelliott 1989).

In the 1970s it was decided that Bournemouth beach should be artificially nourished. In 1974 coarse sand and gravel obtained by dredging the sea floor off the Isle of Wight was shipped in and dumped in a zone 450 m off the Bournemouth shore. On the more sheltered sector west of the pier these deposits were soon washed up on to the beach by wave action, but on the more exposed sector east of the pier shoreward migration did not take place. Instead, the sand and gravel dumped offshore was delivered to the beach through a floating steel pipe by a pump mounted on a pontoon (Fig. 84) (Newman 1976). The compartments between the numerous groynes were thus nourished, using 830 000 m^3 of sand and gravel to form a beach

Figure 84 Beach nourishment in progress at Bournemouth, England, where sand and shingle dredged from the sea floor were pumped on to the shore through a pipe

5 miles long (Willis and Price 1975). Longshore drifting has carried some of the beach material away to the east (Willmington 1983), and the Bournemouth beach has subsequently been maintained by dumping more sand and shingle obtained from the sea floor on the shore at intervals. In 1988–89 sand dredged from the Swash Channel, to improve the navigable approach to Poole Harbour at Sandbanks to the west, was used to renourish Bournemouth beach. This was carried out mainly during the winter, so that the improved channel and replenished beach would be available in time for the holiday season (May 1990).

A similar project was carried out at Mentone (Fig. 85), on the north-east coast of Port Phillip Bay, Australia, an area which is much like Bournemouth geologically. Receding cliffs in soft Tertiary sandstones were stabilised by a sea wall and promenade built in 1937–39, and the 1.8 km sandy beach that had been supplied with sediment eroded from the cliffs then diminished. In 1977 the beach was nourished with about 160 000 m^3 of sand dredged from the sea floor and then pumped on to the shore (Fig. 86). The material supplied to the shore was bulldozed to form a beach terrace, initially about 32 m wide, built 2 m above low spring tide level (Fig. 87). The sand was at first dark in colour because of a coating of silt, clay and organic matter, but rain quickly washed this away. There was a problem because the beach had been emplaced across several storm water outfall pipes, and the first heavy rain saw gullies washed out across the

Figure 85 The nourished beach at Mentone, Port Phillip Bay, Australia. Photograph by Neville Rosengren

beach by outflow. This was remedied by extending the pipes seaward to the outer edge of the nourished beach, but these have become more prominent as the outer edge of the beach has been cut back (Fig. 88).

It was expected that, as on other parts of the east coast of Port Phillip Bay, the deposited sand would drift northward in summer and southward in winter (p. 40). Sand that moved northward was expected to build up along the cliffs of the adjacent Beaumaris coast, but it was decided to insert a groyne at the southern end of the nourished beach to prevent losses of sand southward in winter.

The Bournemouth and Mentone projects both stockpiled sediment in the

Figure 86 Sand being pumped on to the shore at Mentone, Port Phillip Bay, Australia to form the beach shown in Fig. 85

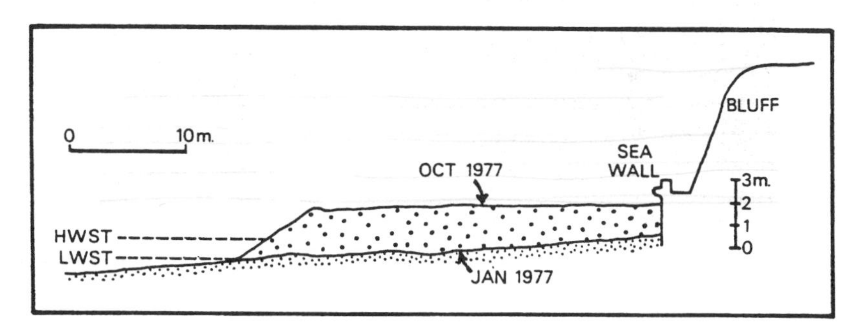

Figure 87 Transverse profile of the beach emplaced at Mentone, Port Phillip Bay, Australia, over a previously depleted beach (see Fig. 104)

nearshore zone for pumping in to the beach. There are limits to the distance over which pumping of sand is effective. On the Belgian coast 500 000 m^3 of sand dredged from the fishing port at Ostend was pumped on to the shore between Bredene and Klemskerke in 1978 to nourish a depleted beach up to a kilometre to the west, but beyond this it was necessary to pump from another source: coarse shelly sand from the offshore shoals at Stroombank and Kwintebank. Some 8.5 million m^3 of

Figure 88 Pipes inserted in the nourished beach at Mentone, Port Phillip Bay, Australia, to convey land drainage to the sea, exposed as the result of beach recession

sand delivered from these sources permitted beach placement on 8 km of coast near Zeebrugge (Kerckaert et al. 1986).

As a sequel to the Bournemouth project, several other seaside resorts on the south coast of England have had their beaches nourished in front of esplanades. Some have used boulder armouring to reinforce the sea wall, and covered this with imported shingle to form a new beach. At Seaford beach erosion began after the building of the large breakwater at Newhaven in 1845 (Fig. 43, p. 97), which cut off the supply of shingle drifting in from the west. As Seaford beach was depleted, storm waves became increasingly destructive, and a sea wall was built to protect the esplanade, with numerous groynes inserted in the hope of retaining what was left of the shingle (see Fig. 64), which was augmented artificially in 1936. Beach erosion continued despite successive elaborations of the structures. Eventually, in 1987, the shingle beach was renourished by the Southern Water Authority with 1.5 million m^3 gravel dredged from the sea floor off Littlehampton, to the west, dumped a kilometre offshore, pumped on to the shore, and deposited over an armouring of large granite blocks imported from Galicia in Spain. The restored shingle beach was then shaped by bulldozers into a broad terrace with a seaward outer slope (Fig. 89), and a retaining groyne at the eastern end (Nicholls 1990).

A similar technique has been used to restore the depleted shingle beach at Sidmouth in Devon (Fig. 90). Protected by sea walls built in the 19th

Figure 89 The nourished shingle beach at Seaford, south-east England, with a terminal groyne in the distance to prevent losses by longshore drifting to the east

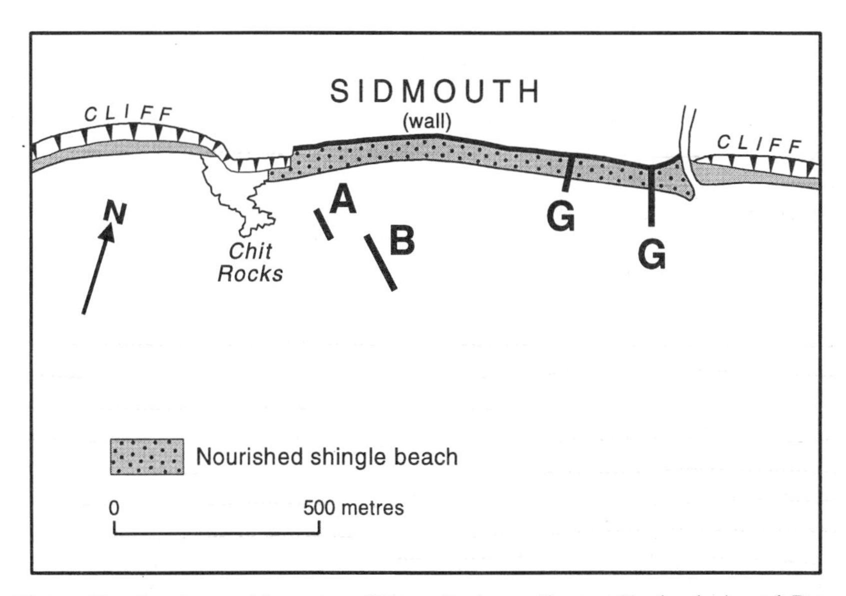

Figure 90 Beach nourishment at Sidmouth, in south-west England. A and B are offshore breakwaters (Fig. 69) to protect the nourished beach from the predominant south-westerly waves. G, groynes

Figure 91 Mobile crane emplacing stone blocks on the shore at Sidmouth, south-west England, prior to the addition of a shingle beach

century, this south-facing seaside resort had been steadily losing its beach, partly because the adjacent cliffs of soft sandstone have continued to recede, so that the esplanade stands slightly forward from the general coastline. In consequence, the beach was more exposed to wave scour, which dispersed the shingle beach alongshore, mainly to the east. It was decided to reinforce the sea wall with a rampart of large stone blocks (Fig. 91), and to cover this with quartzite gravel brought by lorries from an inland quarry in the Triassic Budleigh Salterton Pebble Beds at Woodford, 12 km away: the natural beach had included material originating from coastal outcrops of this formation. The nourished beach was shaped as a terrace with a steep seaward slope (Fig. 92). Two angled offshore breakwaters have been built to protect it from prevailing south-westerly waves and storms, and a terminal groyne (East Pier) to prevent it being lost by longshore drifting to the east. The gravelly fill consisted of well-rounded but poorly sorted pebbles and cobbles, and initially had a pink colour from the earthy Triassic matrix. The beach was hosed down by the local fire brigade in an attempt to get rid of the pink stain, but it is more effectively removed by waves reworking the seaward slope, and is not expected to persist long. The beach restoration, completed in 1995, cost £6.5 million.

The use of coarse material as the basis for a nourished sandy beach was illustrated between Monte Circeo and Terracina, on the west coast of Italy. The beach was nourished in 1980–83, using crushed limestone gravel to

Figure 92 Nourished beach at Sidmouth, south-west England, with gravelly material placed over the boulders shown in Fig. 91

form a sloping terrace which soon acquired a veneer of inwashed fine to medium sand. The coarse gravel thus provided a matrix for sustained sand accretion, as well as increasing shore protection (Evangelista et al. 1992).

Many seaside resorts have sought to nourish their beaches to improve the seaside environment. Cairns, in north-eastern Australia, has very little natural beach, the esplanade being fronted by a large area of mudflats exposed as the tide ebbs. In 1993 the beach was augmented by bringing 4000 m^3 of medium to coarse sand excavated from a nearby delta, and depositing it between two temporary groynes in order to see whether a beach could be maintained along this muddy seafront (Fig. 93). It may be difficult to keep this beach clean, for it stands behind wide mudflats at low tide, and receives muddy sediment when it is washed by turbid water at high tide.

The placement of an artificial beach where none existed previously occurred on the Algarve coast of Portugal. The shore at Praia da Rocha was cliffed and mainly rocky, with only narrow pocket beaches, but in 1969–70 880 000 m^3 of sand obtained from dredging the River Arade and excavation of the harbour at Portimão was pumped on to the beach, and a further 150 000 m^3 was added from these sources in 1983. This formed an artificial beach 1200 m long and up to 200 m wide in front of the cliffs and provided a beach for the seaside resort. A similar project at Praia dos Três Castelos, to the west, failed when much of the 50–70 m wide beach

Figure 93 The augmented sandy beach on the seafront at Cairns, north Queensland, Australia, fronted by mudflats with mangrove seedlings

emplaced in 1983 disappeared within 5 years, evidently because this sector was more exposed to wave scour than Praia da Rocha (Psuty and Moreira 1990).

Artificial beaches have been added on the seaward side of sea walls, as in the Netherlands, and beside harbour breakwaters, as at Cullen Bay, near Darwin in northern Australia. Here a beach of sand dredged from Darwin Harbour was placed in front of a large boulder breakwater built to enclose a marina in a former bay in 1993. This is a macrotidal coast, and the beach, emplaced around high tide level, has been combed down by wave action on ebbing tides to a wide concave profile.

Large-scale beach nourishment, using river gravels, occurred on the Georgian Black Sea coast in the 1970s. Long steep slopes descend to beaches of sand and gravel, which were originally supplied mainly from rivers draining the Caucasian Mountains, delivered to the coast, and spread along the shore. Extraction of sand and gravel from the natural beaches for building purposes, the damming of rivers, and the building of harbour breakwaters all contributed to beach erosion, and the immediate response was to build sea walls, boulder ramparts, concrete blocks and groynes with submerged breakwaters. When these hard structures failed to maintain a beach-fringed coastline it was decided that beach nourishment should be carried out (Zenkovich and Schwartz 1987). Between 1981 and 1987, 9.2 million m^3 of sediment excavated from river channels and alluvial plains was deposited to nourish six beaches on 47.5 km of coastline between Gagra and Batumi, adding 70 ha of recreational shore. The material was carried along the coast in split-hull barges and unloaded. Some was deposited as groyne-like lobes at right angles to the shore, to be spread laterally by wave action to widen the beach; the rest was dumped in shallow water to be washed by waves on to the beach (Kiknadze et al. 1990). The primary aim was to renourish beaches as a means of shore protection, with the benefit that they have provided an improved recreational resource (Fig. 94).

At Ediz Hook, on the southern shores of the Strait of Juan de Fuca, Washington, United States, a spit 5.6 km long and up to 275 m wide shelters Port Angeles Harbour. It had been supplied naturally with sand and gravel from the west, partly from the Elhwa River and partly from the erosion of cliffs cut in glacial drift, but the damming of that river and the building of sea walls to halt erosion along the cliffed coast reduced the sediment supply, and beach erosion became severe. In 1977–78 rock revetments were built, and it was decided to place gravelly material, quarried from glacial drift deposits west of Elhwa River and brought by truck, on the outer shore of the spit. Supplemented in 1985 by further such nourishment, the spit attained a relatively stable configuration (Galster and Schwartz 1990).

In addition to direct placement of beach fill by pumping sediment in through a pipe, as at Bournemouth, running in barges to the shore to unload gravel on the beach, as on the Black Sea coast, or bringing it in trucks, as at Ediz Hook, there is an alternative known as the rainbow technique, most effective where the tide range is small and wave energy low to moderate. Sediment dredged from the sea floor or from estuaries and inlets along the coast can be shipped and delivered directly to a beach by small draught vessels using the rainbow technique, whereby the ship sends a jet of water and sediment over its bows in an arc on the beach, or into the nearshore zone (Riddell and Young 1992). Nourishment of beaches in south-east Queensland has been facilitated since 1988 by the use of the Port of Brisbane Authority dredge, modified to be able to pump sand out over its bow in this way on to a beach. It was first used to place 50 000 m^3 of sand on an eroded beach at Woorim, on Bribie Island, north of Brisbane, and in 1992 to replenish Golden Beach at Caloundra, to the north, with 70 000 m^3 of sand dredged from nearby Pumicestone Passage. The rainbow technique has also been used to restore beaches in compartments between the numerous groynes on the shore at Felixstowe, on the east coast of England.

Emplacement by longshore drifting

Longshore drifting can be used to carry deposited sediment to where it is required if the direction and rate of drifting are well known. A beach can

Figure 94 The beach at the Georgian Black Sea resort of New Afon disappeared after a sea wall was built (above), but was restored by nourishment (below)

be nourished and maintained by repeated injections of sediment at the updrift end, as at Atlantic City, New Jersey, where sand deposited at the north-eastern end is distributed south-west along the city seafront by wave action. In such situations it may be useful to retain drifting sediment on a seafront sector by inserting groynes.

The idea of depositing a large quantity of sand at a selected point on the beach and allowing wave action to spread it along the shore to nourish beaches downdrift was tested at San Onofre, California, in the 1980s. A 200 000 m^3 sand lobe was deposited, and surveys carried out on six shore profiles up to 9 km downdrift over the following 2 years showed that at first there was beach erosion downdrift because the lobe was acting much like a breakwater, but this gradually came to an end as the sand was distributed alongshore. The apex of the lobe migrated at about 2 m/day, and as it moved the sand lobe diminished rapidly in size, shrinking at the rate of about 50% every 300 days, and becoming asymmetrical, attenuated and narrow as it spread downdrift (Grove et al. 1987).

Nourishment of beaches on Sylt (p. 155) used 770 000 m^3 of sand obtained by shallow dredging and deposited as a large lobe protruding from the shore, and material from this was gradually distributed downdrift by wave action. Its progress was monitored on profiles surveyed at 500 m intervals, which showed that after 5 years more than 60% of the sand deposited in the lobe had moved on to the beach downdrift. On the basis of this experience it was decided that the optimum site for lobe deposition should be 1 km updrift of the site chosen initially. The project showed that nourishment by means of redistribution from a deposited lobe was feasible, and that the location of such a lobe should be well updrift of the sector to receive the beach nourishment. A similar principle guided the dumping of urban rubble on the shore near Odessa to nourish beaches downdrift.

If rates of longshore drifting are known, it is possible to maintain a beach by repeated nourishment at an updrift point. Beach erosion on the Gold Coast, in Queensland, Australia, followed the construction of parallel stone breakwaters at the mouth of the Tweed River in New South Wales in 1962, resulting in the trapping of northward drifting sand. As the sand supply diminished, longshore drifting past Point Danger to Coolangatta and the Gold Coast beaches was much reduced, and beach erosion developed. By the 1980s the amount of sand trapped on Letitia Spit, south of the Tweed breakwaters, on the bar off the river mouth, and in the mouth of the river totalled some 7 million m^3 (Pattearson and Pattearson 1983). Beach erosion on the southern Gold Coast beaches by this time also amounted to 7 million m^3 of sand, and it was calculated that net annual longshore sand transport northward along the Gold Coast beaches was about 500 000 m^3/year.

A report by the Delft Hydraulics Laboratory (1970) recommended extensive beach nourishment to restore the eroding beaches, and it was decided to do this by updrift dumping and distribution by longshore drifting. In 1978 sand dredged from the Tweed estuary was dumped on the shore between Kirra and Currumbin (Chapman 1978), but it was not until 1985 that a major beach nourishment programme began, using relatively coarse sand dredged from offshore. It was then calculated that 2 000 000 m^3 of sand was needed in order to restore the southern Gold Coast beaches to their pre-1970 dimensions, and to stabilise them would require addition of another 500 000 m^3/year at the southern end, until natural sand drift of this amount could be revived northward from the Tweed Heads breakwaters with a bypassing scheme (Robinson 1993).

Several beach nourishment projects have used the principle that longshore drifting interrupted by a tidal inlet with strong transverse ebb and flow currents (which act like a breakwater) can be restored by sealing off the inlet or cutting a new one. This has been illustrated on Seabrook Island, South Carolina, where beach erosion became severe when southward drifting of sand was impeded by Captain Sams Inlet, with interception on the northern (updrift) side. In due course the sand accumulating on the updrift side began to form a spit, which grew southward, deflecting the mouth of the inlet, and it was decided to cut a channel through this, and allow 170 000 km^3 of sand to drift on southward. The released sand soon sealed the former inlet, and moved on to renourish the previously eroding beach, widening it by more than 300 m over the next 6 years, south of the new artificial cut (Kana 1989).

The importance of aspect in determining patterns of longshore drifting (see Fig. 12) has influenced beach management projects on the north-eastern coast of Port Phillip Bay, Australia. Seasonally alternating longshore drifting here (p. 40) is seen as a dominance of northward drifting in the summer half-year (November to April) and southward drifting in the winter half-year (May to October), a sequence that is illustrated on Black Rock beach (see Fig. 14). Beach nourishment projects must be designed to allow for this seasonal alternation, and the possibility that the gains at each end of a beach compartment will result in losses past headlands or breakwaters. In these circumstances, coastline orientation in relation to wave climate determines whether an emplaced beach will remain in position, or lose sediment in one direction or the other, so that there is a contrast between predominantly northward drifting on Hampton Beach (facing south-west) and predominantly south-eastward drifting on Quiet Corner Beach (which has a more southerly aspect) a few kilometres away (Bird 1991).

It may be necessary to build terminal groynes to retain a nourished beach within the required sector.

Artificial structures such as marinas may also act as traps for seasonally drifting sediment, as at Sandringham Harbour on this coast. The adjacent beach compartment, between Green Point and Picnic Point, has been depleted following the building of the boat harbour (see Fig. 46), with construction and elaboration of a large boulder breakwater at Picnic Point (see Fig. 24). As on the other beaches on the north-east coast of Port Phillip Bay, there is northward drifting of beach sand between November and April, and southward drifting from May to October. After the harbour was completed in 1954 the breakwater acted as a trap for beach sand drifting southwards each winter, and prevented it from being carried back by south-westerly waves in the summer months. By 1960, little beach sand was left on the Hampton coast, much of it having drifted into the lee of the Sandringham Harbour breakwater to accumulate as a wide prograded sandy plain (Bird 1993a).

The Victorian Division of Ports and Harbors then decided to nourish the northern part of the beach. In order to prevent sand drifting south in the winter months into Sandringham Harbour, resulting in further shallowing and shrinkage of the boat harbour, a boulder groyne 160 m long was constructed in the middle of Hampton Bay. It was built at an angle of 65° to the coastline, so that the emplaced beach was exposed to south-westerly wave action, which could drive the accumulated sand back northward in summer. Early in 1987, 108 000 m^3 of coarse sand was dredged from an area 2 km offshore and piped in to form a beach 40 m wide and 1100 m long, extending from this groyne north to Green Point, now known as South Brighton beach (Fig. 95); it was initially formed as a terrace 2 m above low spring tide level, the top of the groyne being 0.3 m higher.

In the winter of 1987 some of the sand was washed southward over the groyne (Fig. 96), and in the following summer the beach towards Green Point was widened by northward drifting, with some sand moving past this headland. Over the next few years this sequence was repeated, and successive spits built out north from Green Point in summer were pushed back by storm waves in the following winter and added to the next beach. Eventually the losses of sand depleted the nourished beach until it no longer prograded sufficiently in summer to leak past Green Point, and by 1994 it had become fairly stable, apart from continuing seasonal alternations within the beach compartment. There is now discussion on whether a beach can be nourished south of the boulder groyne, and if it is, how emplaced sand can be prevented from drifting into Sandringham Harbour during the winter. The various problems have shown that it is necessary to take account not only of natural beach processes, but also of the impacts of structures such as sea walls and harbour breakwaters on the beach system, when planning beach nourishment projects.

Bypassing

Bypassing is the passage of sediment past an obstacle such as a tidal inlet or breakwater to replenish a beach downdrift. It occurs naturally, but can also be developed artificially as a means of downdrift beach replenishment. It has been used at several places on the coasts of the United States. An early example of beach nourishment using bypassing as a means of artificially restoring the longshore drifting regime was at South Lake Worth Inlet, Florida. After this tidal inlet was stabilised by breakwater construction in

Figure 95 Air photograph of South Brighton beach placed between the New Street groyne and Green Point, Brighton, in 1988. A small area of sand (A) spilled southward over the groyne as the result of southward drifting during the winter, and northward drifting in summer carried sand past Green Point to build a spit (B) in the bay to the north. Photograph taken by the Port of Melbourne Authority. Crown Copyright Reserved

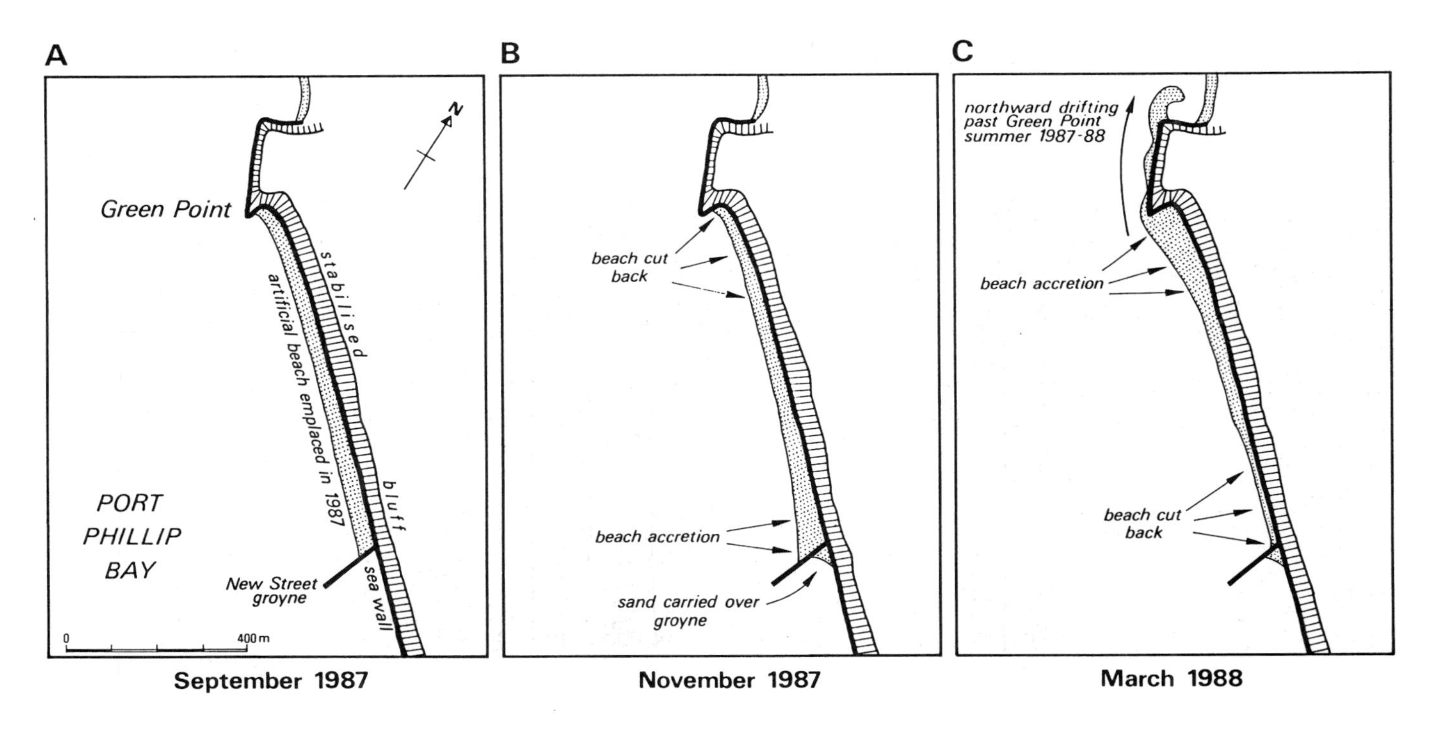

Figure 96 Emplacement of South Brighton beach between Green Point and the New Street groyne in 1987 was followed by southward movement of sand over the groyne during winter and northward drifting past Green Point to build a spit into the bay to the north in the following summer

the 1930s to improve navigability, the nearby beaches soon showed updrift accretion and downdrift erosion (p. 61). When groynes failed to control the downdrift erosion a sand bypassing scheme was introduced, taking about 48 000 m^3 of sand per year from the accreting southern beach round to nourish the depleted northern beach. Similar projects have been developed at Santa Barbara, Ventura, and Channel Islands Harbour in California, both of which had accretion updrift and erosion downdrift of harbour breakwaters.

At Durban, in South Africa, port development alongside an inlet at the southern end of a sandy bay included the building of bordering breakwaters which were extended seaward in 1952 to maintain a navigable approach through the sand bar. Erosion ensued on the beaches in Durban Bight, to the north of the breakwaters, and in 1953 Durban City Council began to pump sand northward under the harbour entrance to replenish them. The sand supply available for such bypassing proved to be insufficient, and so sand dredged from the harbour was dumped in a sand trap, from where it was pumped on to the beaches. These are now maintained with an annual replenishment of 100 000 m^3 (Laubscher et al. 1990).

At the northern end of the Gold Coast, in eastern Australia, Nerang River used to open to the sea through a gap between the Southport spit and South Stradbroke Island, a sandy barrier island to the north. In 1984 a new cut was made through the northern part of Southport spit, and bordered by training walls. It was realised that northward drifting of sand (about 500 000 m^3/year) would cause accumulation on the southern side of these breakwaters, as at Tweed Heads, and that erosion would ensue to the north. To avoid this a sand bypassing scheme was introduced. South of the breakwaters a jetty was built to carry ten pumps at 30 m intervals, which extract the accumulating sand at the rate of 1000 tonnes per hour. This is delivered to a pit, where it is mixed with water from the Nerang River to make a slurry that can be pumped under the entrance and delivered to the beach to the north. A growing spit soon sealed off the old gap, so that a wide beach now extends on to South Stradbroke Island.

Alternatively, sediment can be carried round in trucks, as at Hvidesande in Denmark (p. 152), or ferried past the entrance, as at Port Hueneme, California (p. 183). However, some eroding beaches downdrift of breakwaters have been nourished with sediment from other sources, notably the sea floor, as at Seaford in England, Timaru in New Zealand and Lagos in Nigeria.

The Lagos coastline has been much modified by the building of breakwaters to stabilise the harbour entrance (Fig. 43) and these caused updrift accretion on Lighthouse Beach and downdrift erosion on Victoria Beach (Usoro 1985). To counter erosion, Victoria Beach was nourished with sand pumped in from the sea floor in 1976, but by 1980 waves had washed this away, and were attacking the sea front up to the edge of the built-up area.

Figure 97 The beach on the southern shore of Dungeness, south-east England, before it was widened and raised to protect the nuclear power station

There was renourishment in 1985–86, and it is now accepted that this beach will need to be replenished frequently, perhaps with the aid of a sand bypassing system from the accreting Lighthouse Beach, west of the breakwaters (Ibe et al. 1991).

Recycling

Beach sediment carried along the coast by longshore drifting can be brought back to nourish an eroded beach (Willis and Price 1975). Such recycling (also known as recharging) has been successful at Rye in south-east England and on the Adelaide coast, in both cases with excavation of sediment from a downdrift accretion zone to be taken back by lorries and dumped on the depleted updrift beaches. Shingle dredged from the eastern shore of Dungeness was dumped to maintain the seaward frontage of a nuclear power station on the southern shore (Fig. 97) when there was a possibility that beach erosion could undermine it (Townend and Fleming 1991).

At Rye, the shingle beach is subject to eastward longshore drifting, a process which over the past few thousand years has supplied vast numbers of flint and chert pebbles to build the great cuspate foreland of Dungeness. Stabilisation of the mouth of the River Rother by breakwaters in the 19th century began to intercept this drifting shingle, so that the beach downdrift

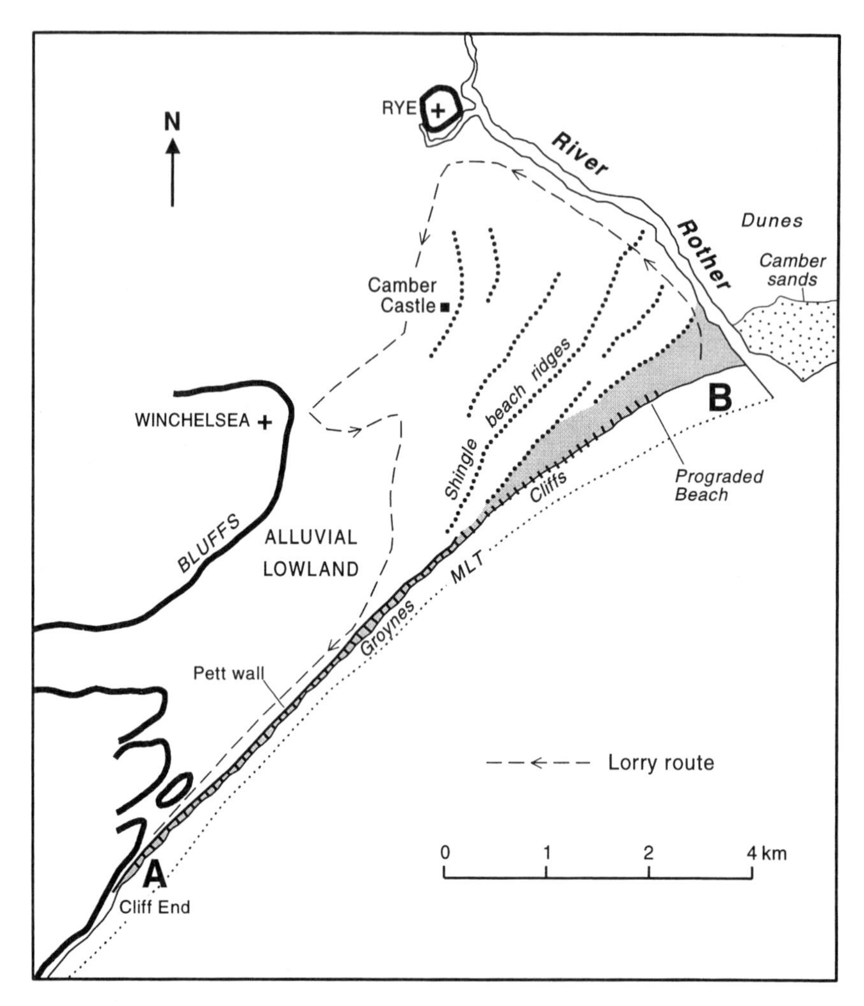

Figure 98 The coast near Rye, in south-east England, showing the diminished beach at Cliff End (A) and the area of progradation alongside the breakwater at the mouth of the River Rother (B), with the route taken by lorries recycling shingle from B to A

at Camber became sandy (Fig. 98). Soon afterwards the beach to the west, at Cliff End, began to erode because of a reduction in the supply passing the Fairlight coast, where recurrent landslides interrupted longshore drifting from Hastings. In 1934 some shingle was taken from the accreting area adjacent to the Rye breakwater and returned to the depleted beach to the west, using a beach railway. In 1955 the Kent River Board began regular ferrying of lorry loads of shingle taken from alongside the breakwater round to Cliff End and then dumped them to restore the depleted beach

Figure 99 Shingle placement on the beach at Cliff End (A on Fig. 98), brought by lorry from the prograded beach downdrift, at the mouth of the River Rother (R)

(Fig. 99): in the first year 155 000 m^3 of shingle was returned to the western end of the beach, and in subsequent years similar quantities were recycled. In 1979, for instance, more than 100 000 m^3 of shingle was delivered in this way to four sectors of beach near Cliff End (Eddison 1983).

Reference has been made to recycling of sand on the beaches in Adelaide, South Australia (p. 141). Sand lost from the southern beaches has drifted northward to Port Adelaide, where the wide beach has been used as a source of sand trucked back southward to the eroding beaches (Wynne 1984). Problems developed because of opposition from people who wanted the wide Port Adelaide beach preserved because of its scientific interest and the natural history of the dune vegetation that had colonised the backshore. There were also protests at the continual traffic of sand-laden lorries through the seaside suburbs of Adelaide, and because of this the project was terminated. The nourished beaches were soon depleted, and an alternative source of sand was sought. Sand is available from inland quarries, but cannot be used because of the truck traffic problem. Sectors of the beach narrowed by erosion, especially during stormy periods, are being renourished with coarse sand dredged off Port Stanvac and dumped inshore, to be pumped on to the beach.

Lorries were also used to recycle beach sand that had drifted away along the shore at Morib in Malaysia (p. 149). Recycling could be carried out by barges travelling along the shore, or by pumping the sand back from the updrift end of the beach through a pipe, as at Narva Bay in Estonia. Recycling has the advantage that the shape and size characteristics of the material taken from downdrift will be similar to those on the depleted updrift beach, but there will be gradual attrition, and supplementary coarser material may need to be imported eventually.

Use of groynes

Groynes can be useful in retaining a nourished beach placed on a coast dominated by longshore drifting. In Monterey Bay, California, sand from a quarry was dumped on the shore alongside a retaining groyne built at Capitola to prevent losses downdrift (Griggs 1990). Reference has been made to groynes built to retain nourished beaches at Seaford and Hengistbury Head in England, Wrightsville in North Carolina, USA, and Mentone, Brighton and Sandringham in Port Phillip Bay, Australia.

When shingle dredged from the sea floor was brought in to nourish a depleted beach in front of the sea wall at Lodmoor, east of the seaside resort at Weymouth, in 1995, there was a possibility that during south-easterly storms shingle from the augmented beach would drift westward on to the sandy resort beach. In order to prevent this a T-shaped retaining groyne was constructed at Melcombe Regis, at the western end of the nourished beach.

The disadvantage of groynes is that they generally result in erosion downdrift. On Sandy Hook, New Jersey, the shore had been protected by an 11 km sea wall, to which groynes were added, but erosion beyond the end of these structures began to cut out a bay. In 1977, 152 920 m^3 of sand was trucked in and deposited in the bay to provide protection from storm damage, but it soon drifted away northward (Nordstrom et al. 1979). It would have been possible to go on extending the groyne field downdrift, but Nordstrom and Allen (1980) suggested it might be preferable to abandon the groyne field and provide a supply of sand at the updrift end sufficient to maintain a protective beach along the coast by longshore drifting.

Groynes are of little use where sediment from the nourished beach is being withdrawn to the sea floor, as at Virginia Beach on the Atlantic coast of the United States (p. 192). In such conditions it may be possible to retain a nourished beach by building nearshore underwater breakwaters to prevent seaward losses, as at Niigaata in Japan (Chill et al. 1989).

Use of shoreward drifting

On swash-dominated beaches there is the possibility of dumping sediment in the shallow nearshore area and allowing it to wash on to the beach. Studies of onshore–offshore movements of sediment in relation to wave and current regimes (usually seaward movement during storms and shoreward movement during calmer weather) are necessary to determine the depth from which such shoreward transportation will occur, and therefore the optimum location for dumping material to be washed on to the beach.

At New River Inlet, North Carolina, an attempt was made to nourish an eroding beach by dumping sand on the nearshore sea floor in the expectation that it would be washed on to the beach by wave action. Some 26 750 m^3 of coarse sand dredged from New River was dumped on the sea floor by split-hull barges, and its movement was followed by monitoring beach and nearshore profiles. The study showed that the sand deposited in depths of less than 4 m moved shoreward over the ensuing 13 weeks, whereas sand deposited at greater depths moved seaward. The sand that moved shoreward was deflected along the shore by obliquely arriving waves to beaches down the coast (Schwartz and Musialawski 1977). The project indicated that if sufficient sand is deposited in the nearshore zone it will move shoreward from a specific depth, but that longshore drifting will determine which sectors of the beach receive it.

Sand placed on offshore bars has moved in to beaches on the Pacific coast of the United States (T.D. Clayton 1989), and on the Gold Coast in Australia. At Burleigh Heads, 100 000 m^3 of sand was dredged from the sea floor 1.5 km offshore (where the water is 18–25 m deep) in 1985, brought in and dumped on sand bars in the nearshore zone. Within a year much of this sand had been washed on to the beach.

The proportion of sand dumped in the nearshore zone that moves up on to the beach varies with wave conditions. In North Carolina, Schwartz and Musialawski (1977) found that up to 75% of the dredged river sediment dumped in the nearshore zone was washed on to the beach, but if stormy conditions followed the nearshore dumping little or none of it moved onshore. There is usually also some longshore drifting, which will determine which sectors of the beach actually receive the inwashed sediment.

Use of nearshore structures

Offshore breakwaters have been used to create a pattern of refracted waves that will concentrate sand deposition and prograde the beach in the lee of the breakwater. This has been illustrated at Santa Monica, California (p. 71). At Port Hueneme, California, a breakwater was built parallel to the coast on the updrift side of the harbour entrance in 1940, and the sandy cusp that formed on the beach landward of it (Fig. 100) has been excavated at the rate of about 400 000 m^3/year by a floating dredge to produce sand which is ferried past the entrance to replenish wasting beaches on the downdrift shore (Johnson 1959).

It has been suggested that a floating breakwater, anchored off successive sectors of the shore, could be used to induce local accretion of sand and

Figure 100 The Californian coast at Port Hueneme, where an offshore breakwater induces cuspate accretion, and sand from this is ferried across to nourish the depleted beach to the south (left)

gravel by shoreward drifting of sediment to nourish a beach in stages along the coast.

Reference has been made to the use of submarine breakwaters to diminish wave scour and protect a nourished beach at Niigaata, Japan (p. 128). At Marina di Cecina, on the Tuscan coast, Italy, beach nourishment (18 500 m^3) was accompanied by the building of retentive groynes with undersea extensions. These have been successful, but two of the submerged breakwaters caused current scour leading to a sediment deficit and some beach erosion, and these are being allowed to disintegrate (Ciprani et al. 1992).

Backpassing and beach re-shaping

Losses of sediment seaward from a beach, particularly during stormy phases, can be offset by backpassing, the retrieval of beach material that has been swept offshore and its return to the beach. This is analogous to recycling of beach material carried away by longshore drifting, mentioned above, and is important in beach profile nourishment, which is discussed below.

Backpassing is a possibility where wave energy is low, especially if there is a wide intertidal area, as at Rosebud in south-eastern Australia. On the southern shores of Port Phillip Bay a long, gently curving sandy beach, much used by summer holidaymakers, extends from Rosebud to Rye. It is fronted by multiple parallel sand bars that run out seaward across a shallow intertidal area up to 500 m wide, and move to and fro in response to alternations of obliquely arriving wave action. In the 1950s erosion of the beach prompted the building of sea walls and groynes, but depletion continued, and in 1963 it was decided to renourish one sector by bulldozing sand in from the nearshore sand bars at low tide. This was successful, and during the next 20 years several sectors of the beach were built up and widened in this way. Between 3000 and 5500 m^3 of nearshore sand were delivered to the beach annually, and parts of the nearshore area deepened by up to 30 cm as the bulldozer scooped sand shoreward.

However, there was an ensuing problem of dense seagrass infestation in the bulldozed areas, and the beach and nearshore area had to be restored by nourishment. This was achieved in 1985 by dumping a series of artificial transverse bars of fine sand 5 m wide and 120 m long, spaced at 100 m intervals (Fig. 101), to be widened and moved to and fro by wave action until they buried and destroyed the nearshore seagrass beds. Within 2 years the artificial transverse bars had been re-shaped by wave action, and their lateral migration had reduced the seagrass area to a few small patches amid the distributed sand (Parry and Collett 1985).

Shingle beaches on the coast of England and Wales have been restored by bulldozing gravel up to the back of the shore. The combing of shingle down the beach by plunging waves during storms is thus countered by artificially restoring the upper beach profile in front of a sea wall or eroding cliff. This has been effective at Dunwich, East Anglia (Fig. 102), as a short-term procedure, where the aim is to increase upper beach protection of soft cliffs cut in glacial drift. Where storm downcombing becomes frequent it may be necessary to add more sediment in order to nourish a higher and wider beach that is more protective.

After storm surge flooding overtopped Chesil Beach, near Portland, in the 1970s the beach crest was raised by bulldozing up shingle, and then stabilised with a capping of gabion mattresses (Fig. 103). Shingle has also been bulldozed up on beaches along the south-western shores of Dungeness, in south-east England. Where beaches lose material by overwashing during storms it can be bulldozed back from the inner slope to the beach face. This has been successful on the shingle barrier beach north of Timaru, New Zealand, where material washed over into Washdyke Lagoon by storm waves was retrieved in this way (Kirk and Weaver 1985).

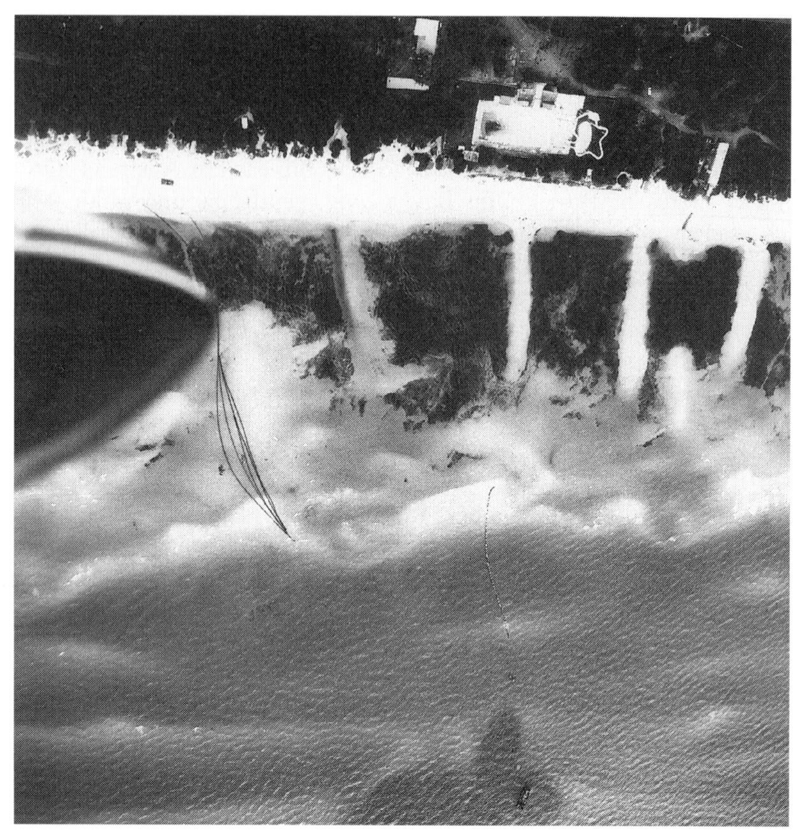

Figure 101 The use of artificial sand groynes at Rosebud, Port Phillip Bay, Australia, to restore the nearshore bar profiles in an area where seagrass (seen as dark areas) had invaded hollows previously scraped by bulldozing sand bars up on to the beach. Wave action soon redistributed the sand groynes to restore the natural topography. Photograph by Neville Rosengren

A different kind of backpassing may be necessary where wind action carries sand to the back of the beach. At Harrison County, Mississippi, sand blown from the nourished beach by occasional strong wind action during hurricanes piles up against the wall to the rear, and periodically has to be taken back by trucking to restore the beach profile.

Many seaside resorts improve their beaches for summer holiday use by sweeping back sand and shingle dumped on their esplanades by winter storms. Some, such as Weymouth, take care to maintain a clean, flat sandy beach as an attraction for children and a venue for sand castle competitions each summer. Each winter, sand is washed and blown round the shore of Weymouth Bay to accumulate as low dunes in front of the sea wall at the southern end of the esplanade, and in spring this is collected and redistributed across the beach.

Figure 102 Shingle on the beach at Dunwich, eastern England, has been bulldozed up to form a protective ramp at the base of receding cliffs

Figure 103 After the south-eastern end of Chesil Beach, southern England, was overwashed by a storm surge the ridge crest was artificially raised and protected by dumping gabion mattresses of caged shingle

Overfill

It is generally necessary to add more beach material than is necessary to restore a beach to its natural dimensions in order to allow for expected losses onshore, offshore or alongshore. As has been noted, Dutch beaches are usually overfilled by about 20%. James (1975) defined the Overfill Ratio as the volume of material necessary to restore a beach similar to the natural beach, allowing for losses of sediment until the grain size characteristics had been sorted to match the natural distribution. The Renourishment Factor is the ratio of the rate of erosion of beach fill material to the preceding rate of natural beach erosion, indicating the frequency of replenishment necessary to maintain a stable beach volume. Overfill is also necessary to anticipate losses due to spilling of sediment out of the nourished area, around a terminal groyne or bordering a headland or terminal groyne, as at Hampton Beach in Port Phillip Bay.

Beach configuration and stability

There have been suggestions that once a beach attains a particular shape in plan it will become stable (p. 53). For example, beaches shaped by obliquely arriving waves alongside a headland or breakwater develop an asymmetrical curvature sometimes known as crenulate, 'half-heart' or 'zeta-curve' configurations. The notion that 'headland breakwaters' can be used to shape stable nourished beaches within intervening compartments by attaining such a configuration, related to the refraction of obliquely arriving waves, is based on the work of Silvester (1976), who indicated that beach stability (in the sense of zero longshore drifting) could be attained when bays assumed this configuration. Intermittent breakwaters were constructed on the shore of East Coast Park in Singapore, where it was found that longshore drifting diminished as the intervening beaches attained a crenulate shape, but erosion has continued on the asymmetrical beaches formed in this way, so that the problem of coastline instability remains.

Nevertheless, some beach outlines are more stable than others, and a nourished beach will be more persistent if it is placed on an alignment that is compatible with incident wave regimes. Reference has been made (p. 174) to the importance of aspect in relation to prevailing wave patterns in determining directions of longshore drifting, notably on the north-east coast of Port Phillip Bay, Australia. The case of Wrightsville, North Carolina, has been mentioned (p. 148) as an example of initially incorrect alignment of a nourished beach, with increasing stability as the beach became realigned more closely to the pattern of incoming waves. In these examples, use was made of breakwaters or terminal groynes to delimit a nourished beach in such a way that it would be correctly aligned. Attention to dominant wave regimes and patterns of wave refraction approaching a coastal sector can improve a beach nourishment project by selecting a suitable alignment for the project design.

Shore profile nourishment

Similar considerations apply to the stability of beaches with a particular shape in profile. Most beach nourishment projects form a beach terrace, which is then re-shaped by waves and currents towards a natural concave profile, often with sand bars just offshore. It has been suggested that it may be more useful to nourish the whole profile, including backshore dunes and the nearshore sea floor, and not just the upper beach terrace, in order to attain a more stable transverse configuration (Bruun 1990). Shore profile nourishment is preferred to upper beach nourishment because the latter leaves unnaturally steep seaward edges, which can be reflective and a cause of nearshore scour, and are also subject to erosion and re-shaping, with often rapid initial losses. Profile nourishment reduces these losses, and has lower construction and maintenance costs. It also permits the use of a wider range of grain sizes, subject to sorting by wave processes into appropriate zones on the profile.

The aim is to establish a relatively stable 'equilibrium profile' of the kind discussed in Chapter 2 (p. 53). As there noted, beaches that show concave profiles are more stable than those with straight, convex or irregular profiles, and once concave profiles are attained they become relatively (but not absolutely) stable. The gradient of the concave profiles varies with grain size and preceding wave conditions. Subsequent oscillations occur with episodes of storm wave erosion and fine weather accretion, and in The Netherlands the value of backshore dunes as a reservoir of sand and a barrier to storm waves and marine flooding has long been realised. On the Atlantic coast of the United States, Kana and Stevens (1990) discussed techniques of beach and dune profile restoration following erosion by a hurricane. Hansen and Byrnes (1991) proposed a beach profile change numerical model to optimise the nourishment and maintenance of shore profiles on nourished beaches.

Bruun (1990) also noted that shore profile nourishment required dredging and dumping equipment of the kind used in The Netherlands (Stive et al. 1991). A transverse profile can be maintained by backpassing (p. 184), using permanent offshore dredging and pumping stations, but this could lead to frequent disruption of sea floor plant and animal communities and fish habitats, and some would consider the offshore structures obtrusive.

Renourishment of the whole of the shore profile was carried out after the failure of several beach nourishment projects which dealt only with the upper beach terrace at Ocean City, New Jersey. Detailed investigations prepared the way for a project which used 4.6 million m^3 of sediment to shape a beach 30 m wide, with a concave shore profile on the seaward side, as well as a backshore dune (Fulford and Grosskopf 1989, Anders and Hansen 1990). Subsequent changes were monitored, and in January 1992 a major storm removed most of the beach and part of the dune (Houston 1991b), but there was little property damage in the resort, and it seemed likely that a protective beach and dune could be maintained if the beach profile were episodically renourished after each storm. Nourishment of the whole of the shore profile, including the nearshore and backshore zones, was thus seen as a more effective way of establishing and maintaining a protective beach than simply dumping sand to form an upper beach terrace.

At Miami, the beach nourished in 1975 included a low backshore dune, a flat terrace and a gentle seaward slope, a landform association that has remained fairly stable, proving remarkably resilient even during Hurricane David in 1979 (Finkl 1981). In The Netherlands engineers have sought to maintain beach and dune systems that protect extensive areas of polder land below high tide level by nourishing the nearshore zone as well as the beach, and providing sufficient sand to maintain the dunes behind the beach (Roelvink 1989). Fifty such projects, using a total of 60 million m^3 of fill, were completed between 1952 and 1989, most of them since 1970 (Roelse 1990). Much of the Dutch coast is subsiding, and it has been calculated that between 6 and 10 million m^3 of sediment fill will be required annually to maintain the existing coastline (Louisse and Kuik 1990). Where necessary the coast will be built forward to reduce impacts of future erosion, bearing in mind the probability of an accelerating sea level rise in the next few decades (Pluijm 1990).

If a sufficiently wide beach is formed, wind and wave action will then shape a shore profile that includes backshore dunes as well as nearshore sand bars. At Noosa in Queensland, Australia, the beach was nourished in 1988, some years after erosion had developed following diversion of the Noosa River outlet westward to protect a holiday estate on low-lying sandy Hays Island from river scour (Coughlan 1989). Sand sprayed by the rainbow technique on to the shore in Granite Bay and Tea Tree Bay drifted round to nourish the main beach at Noosa, where it was retained by a groyne built out from the eastern side of Noosa River. In due course it prograded to form a wide area of bare sandy beach, the landward part of which was shaped by wind action into backshore dunes, now stabilised by the planting of trees and shrubs.

Changes after beach nourishment

Changes are to be expected on beaches that have been nourished. Such beaches will be eroded for the same reasons that the preceding natural beaches were depleted (Riddell and Young 1992). Most nourished beaches begin to lose sediment as soon as they have been emplaced, some of the beach material being washed or blown away alongshore, some being swept to the backshore and beyond, and some withdrawn to the sea floor. Monitoring and mapping are necessary to determine the rates and patterns of such losses, when and how much supplementary beach material is required and where it should be placed (Foxley and Shave 1983). Changes are usually measured by making repeated surveys along transverse profiles from the back of the nourished beach out on to the nearshore sea floor, and linking these by alongshore surveys. Supporting evidence is usually available from sequential air photographs. The aim of monitoring is to understand the processes that erode and distribute emplaced beach material. Such information can guide future beach management, such as the insertion of groynes, the introduction of regular renourishment updrift on beaches that are losing sediment alongshore, or the need to repeatedly restore the profile of a beach that is losing sediment offshore.

Long-term monitoring of several nourished beaches on the Atlantic coast of the United States showed rapid initial losses followed by more gradual depletion. At Long Beach Island, New Jersey, for example, the beach nourished in 1979 eroded rapidly for the first 18 months, then more slowly, until by 1986 all of the sand that had been added had disappeared, and the beach had returned to its pre-1979 width and profile.

As changes proceed, the plan and profile of the beach are re-shaped into patterns more closely adjusted to the prevailing wave and current regimes, and influenced by the grain size characteristics of the remaining sediment (Everts et al. 1974). Usually the finer sediment is removed first, leaving the beach coarser in texture, and often steeper in profile. On the Adelaide coast near Brighton, in South Australia, a beach formed by deposition of sand dredged from the sea floor was initially dissipative, but as the finer sand was removed it became reflective, with a steeper upper beach persisting as finer foreshore sediment was washed away.

On the shores of Port Phillip Bay, Australia, the beach terrace formed by sand nourishment at Mentone in 1977 (Fig. 104) was monitored by repeated surveys of transverse profiles at 30 m intervals. These showed a reduction of beach terrace width from 32 to 22 m in the first year, due to re-shaping of the seaward slope to a slightly concave profile with an average gradient of 1:9. At the same time, parallel bars of fine sand formed in the nearshore zone, partly from the finer fraction of material withdrawn from the restored beach material by storm waves, and partly from the fine silty sands already present on the sea floor. The beach terrace was thereafter cut back about 1 m per year, recession occurring mainly during brief episodes of storm wave activity, especially when these coincided with high tides. In 1984 Mentone Beach was renourished by pumping in a further 18 500 m^3 of coarse sand to restore the profile to its 1977 dimensions (Jones and Schafter 1986). This has been effective, for although there has been further gradual depletion, the beach still had an average width of 20 m in 1995.

Figure 104 The beach at Mentone, Port Phillip Bay, Australia, two months after nourishment in 1977, showing re-shaping of the seaward margin of the beach terrace by wave action. Compare Fig. 88, taken 20 years later

The value of monitoring a nourished beach was well illustrated at Virginia Beach, on the Atlantic coast of the United States. Numerous groynes had been built in the late 1940s in the hope of halting beach erosion, but these failed to retain the beach, and in 1952–53 just over 1 million m^3 of sand, coarser than that on the original beach, was added. Monitoring showed continuing depletion of this beach, and it became clear that the groynes were of little use, because most of the sand was being withdrawn to the sea floor, instead of drifting away alongshore. It is necessary to determine the direction of beach losses before inserting groynes, which are more effective when beach material is moving alongshore rather than seaward. The conclusion was that Virginia Beach could only be maintained by frequent renourishment or backpassing of sand lost to the sea floor. This approach to the replenishment of Virginia Beach illustrates Pilkey's (1990) suggestion that beach nourishment should be regarded as experimental, with improvements based on experience gained from continued monitoring.

Mapping of changes on a nourished beach can determine patterns of movement alongshore. In 1963 sand was dumped on the shore near Absecon Inlet, north-east of Atlantic City, New Jersey. Repeated mapping showed that the beach fill was shaped into a lobe that migrated south-westward along the shore at 2–3 m per day. After 2 years this lobe arrived to augment the beach in the vicinity of the main pier, but it continued to move along the shore in front of the boardwalk, and then beyond, so that the widening of the Atlantic City beach was only temporary. The response in 1970 was to add a further 596 000 m^3 of sand near Absecon Inlet, and this also moved alongshore to the Atlantic City pier, where the beach was widened in 1972, and moved on south-westward (Everts et al. 1974). It was then clear that on this drift-dominated coast a beach could be maintained at Atlantic City by frequent small injections of sand at the north-eastern end (Pilkey and Clayton 1987). Groynes were placed on the shore to reduce the rate of drifting to the south-west in an attempt to keep the beach at Atlantic City, where it was wanted (Weggel and Sorensen 1991).

Monitoring of nourished beaches has generally been restricted to the emplaced sector, to decide when and where further nourishment is necessary, but there should also be mapping and monitoring of changes on adjacent shore and nearshore areas to which eroded sediment may move. There is a risk that sediment from nourished beaches will drift alongshore and accumulate in boat harbours, or as shoals impeding navigation at the mouths of rivers and creeks. On the other hand, it may prove beneficial in nourishing other beaches along the coast. Surveys following the nourishment of the beach at Bournemouth, on the south coast of England, showed that the deposited sand and gravel were drifting eastward along the shore, augmenting beaches as far as Hengistbury Head, where a terminal groyne has intercepted drifting material to form a widened beach, protecting the backing cliffs from further erosion (see Fig. 61) (Nicholls and Webber 1987).

On the north-east coast of Port Phillip Bay, Australia, where longshore drifting shows a seasonal alternation (p. 40), surveys showed that sand lost from the nourished beach at Quiet Corner, Port Phillip Bay (Figs. 105 and 106), was carried south-eastward each winter by longshore drifting by waves arriving from the west, so that as the nourished beach became narrower the beaches to Banksia Point and beyond were widened (Fig. 107). Because of the southerly aspect of this beach, there was little north-westward drifting in the summer, so the emplaced beach did not grow in that direction (X on Fig. 106, C). Successive profile surveys on Quiet Corner Beach, Port Phillip Bay, showed that the emplaced beach terrace was also cut back by storm waves. Some of the finer sand withdrawn from the beach was deposited as a sand bar that persisted in the nearshore zone, the crest of which moved shoreward in calmer weather and seaward after storms (Fig. 108).

Monitoring can determine the quantities of sediment lost from a beach sector. At Kirra, at the southern end of the Gold Coast in Queensland, Australia, over 5 million m^3 of sand was dumped on the beach in several phases between 1985 and 1990. A survey in May 1992 showed that 87% of this nourished material was still on this beach or in the nearshore region,

Figure 105 A view over the nourished beach south-east of Quiet Corner, Port Phillip Bay, Australia. Photograph by the Port of Melbourne Authority

the remainder having drifted alongshore to augment beaches to the north (Delft Hydraulics Laboratory 1992).

Decisions on when and where a nourished beach should be replenished, and how much beach fill is required, can be made in terms of information from such mapping and monitoring. There were rapid changes after renourishment on the beach at Wrightsville, North Carolina, monitoring of 50 transverse profiles showing rapid initial losses, some 66% of the renourished sediment being lost within the first year. The erosion rate slackened as the beach profile, originally a terrace with a convex seaward slope, became

Figure 106 (*opposite*) The coastline south-east of Quiet Corner, Port Phillip Bay, Australia. (A) In 1980, after sea wall construction and cliff stabilisation was followed by beach depletion. (B) In 1984, when an artificial beach was emplaced. (C) In 1989, when the beach had been narrowed by erosion, with a sand bar which had formed offshore (see Fig. 108), and south-eastward drifting of sand to Banksia Point

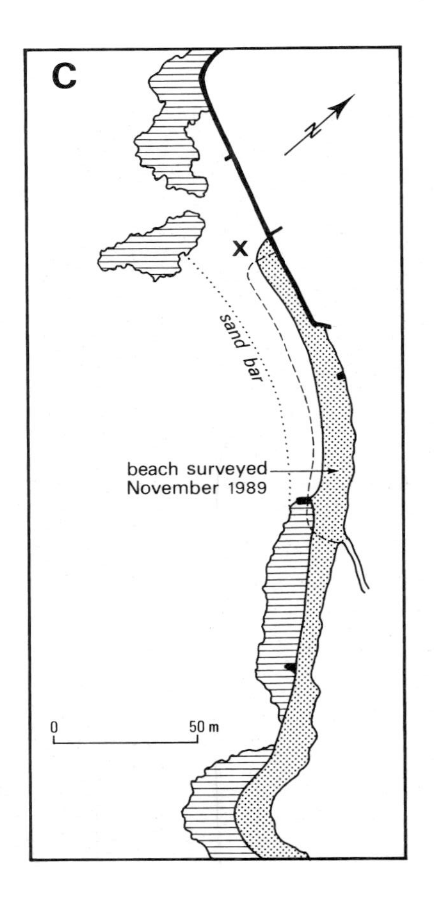

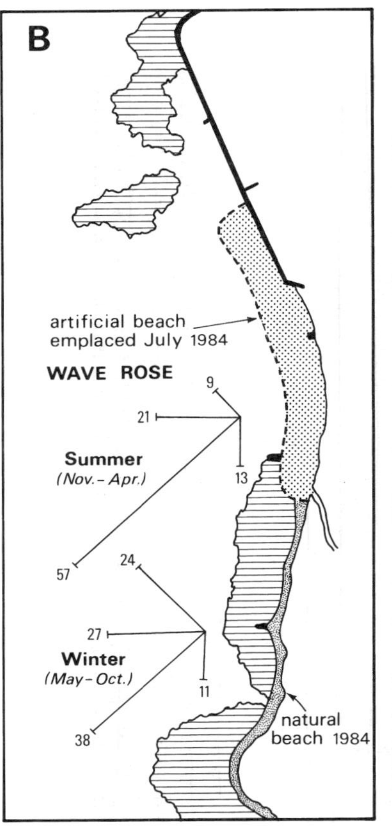

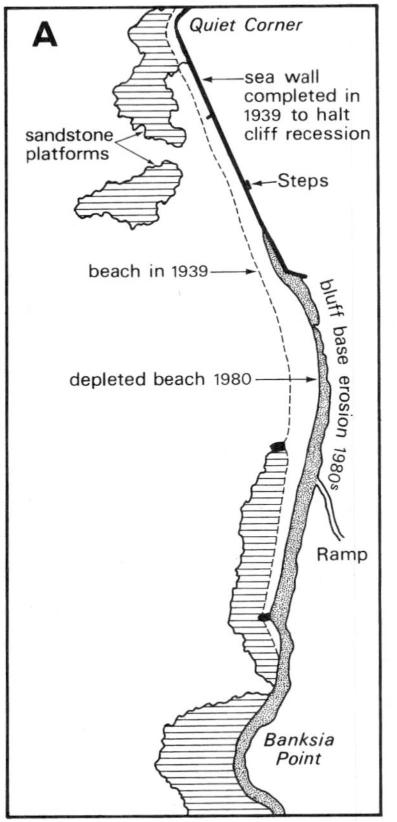

Figure 106 *For caption see opposite*

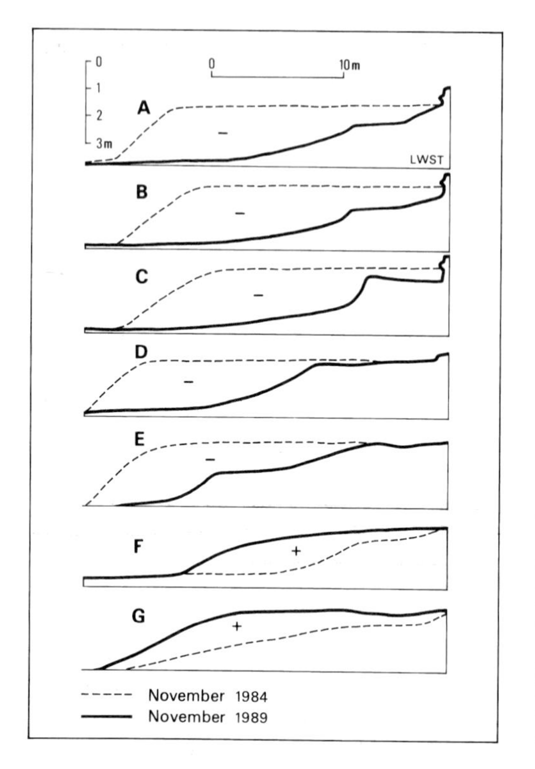

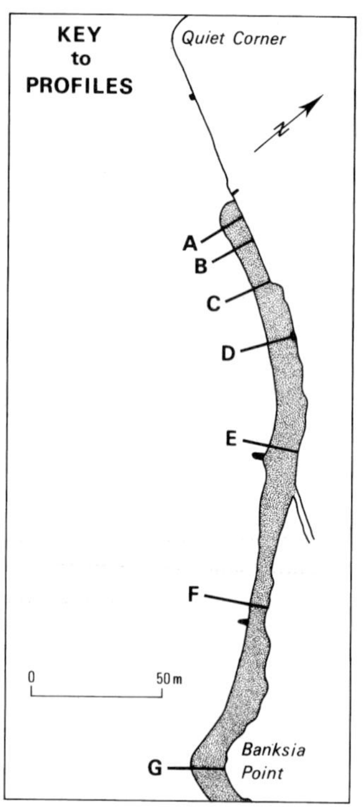

Figure 107 Changes on profiles of the nourished beach south-east of Quiet Corner, Port Phillip Bay, Australia, between 1984 and 1989, when it was depleted by south-eastward drifting of sand to Banksia Point, where the beach was built upward and outward

concave, and there was then more gradual recession, the beach maintaining a more or less constant seaward slope. The initial rate of erosion on the nourished beach was initially ten times that of the preceding natural beach erosion (Pilkey and Clayton 1987), the loss rate eventually declining to the long-term natural erosion rate after 8 years. This was the effective residence time of the beach fill, and indicated that if a nourished beach was to be maintained here it would need to be renourished at intervals of about 5 years.

Another example of monitoring being used to plan frequency of renourishment came from Delray Beach, Florida, where erosion was extensive between 1950 and 1970, and revetments built to preserve the coast road had to be frequently repaired. In 1973 it was decided to nourish the beach, using sand dredged from about a kilometre offshore, and over 1.5 million m^3 of sand was pumped in to widen the beach by 30 m. This protected the coast road and seaside resort, and improved the beach for recreation, but

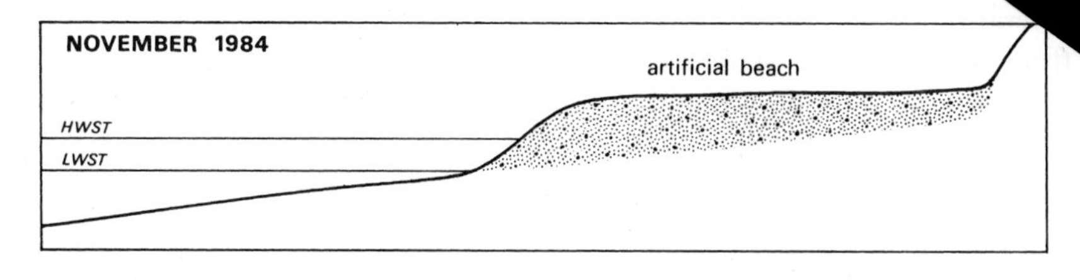

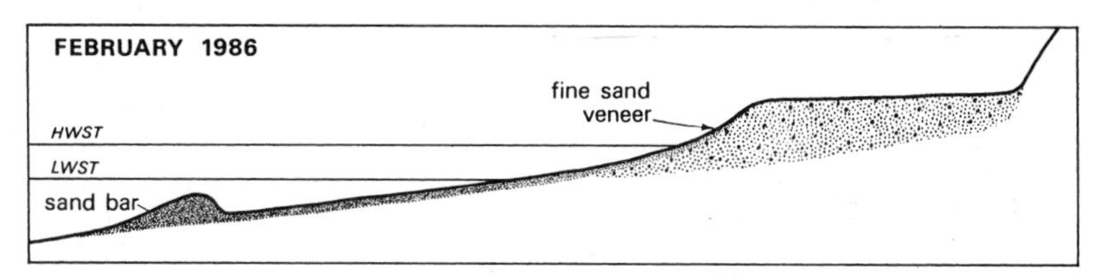

Figure 108 After the emplacement of a beach as a backshore terrace south-east of Quiet Corner, some of the sand removed by storm waves was deposited as a nearshore sand bar. Gentler wave action during calmer periods then added a veneer of fine sand to the concave beach profile

monitoring showed that it was losing about 120 000 m^3 per year. This calculation was the basis for a 5-yearly renourishment programme of restoring the beach with 600 000 m^3 of sand (Campbell and Spadoni 1982).

Management of the Adelaide beaches in South Australia has also used monitoring of changes in beach width and profile to determine where and how much sand replenishment was needed (Fotheringham and Goodwins 1990). Instrumental surveys have been carried out since 1975 on a series of beach profiles spaced approximately 750 m apart, with closer monitoring where necessary at 50 m intervals. The results are presented in the form of maps that shade areas with surface gains or losses of between 0.2 and 1.0 m, and more than 1.0 m. More recently the data have been processed using Geographical Information Systems to produce coloured contour maps of the beaches, from which patterns of gain and loss can be identified, and areas of developing deficit replenished by dumping sand (Noyce 1993).

Beach management on the shores of the Great Lakes has faced the problem of changes on nourished beaches accompanying irregular oscillations in water level of up to 5 m over periods of several years. When lake levels rise beach erosion is prevalent, and when they fall there is progradation. In 1974, 175 000 m^3 of beach fill was emplaced on the shore of Michigan City, Lake Michigan, and a further 61 000 m^3 in 1981. Monitoring showed that the emplaced beach profile was soon modified by wave action, becoming relatively stable in relation to variable lake levels (Jansen 1985). Nourished beach profiles can thus adapt to hydrodynamic variations, often with a time lag of several weeks or months (Thompson 1987). The US Army Corps of Engineers has since nourished beaches at the Lake Michigan coast, using coarse sand dredged below the 5 m contour. Most of these beaches have on even during phases of high lake level, becoming term oscillations in lake level (Macintosh and Anglin

Bea... ...nent for coast protection

The traditional response to coastal erosion has been to build solid structures, such as sea walls or boulder ramparts, to protect the coastline, but it has been realised that a nourished beach which prevents storm waves from attacking the base of a cliff can be as effective a means of coastal protection as solid structures, providing it persists for a sufficient period to be cost-effective.

Nourished beaches have generally been added in front of previously built sea walls, or inserted between groynes, as a supplementary means of coastline protection, to make the coastline less artificial in appearance and to provide a recreational resource. Addition of a nourished beach on the seaward side of a sea wall has been seen as a way of 'softening' hard engineering at several sites on The Netherlands coast, and at Melaka in Malaysia. When land was reclaimed from the sea to create Montego Freeport, Jamaica, sand dredged from the harbour area was used to nourish beaches on the seaward edge of the reclaimed area (Jones 1975).

In Japan coastal erosion has been countered by extensive sea walls and tetrapods, built because the government requires that the coastline be stabilised by hard structures as a 'permanent' solution (Nakayama et al. 1982). Until the late 1950s the beaches of Tokyo Bay were a major recreational area, but land reclamation and the spread of port and industrial facilities in the 1960s overran them to produce a coastline dominated by concrete sea walls. Beach nourishment in front of these walls began in the 1970s, using sand dredged from the bay floor, and by 1990 nine artificial beaches with a total length of 13 km had been formed as part of a series of intensively used coastal recreation parks (Koike 1990) (Fig. 109).

In recent years there have been several projects using nourished beaches as an alternative to engineering works to prevent further cliff recession. On the north-east coast of Port Phillip Bay, Australia, a sector of natural vegetated bluffs south from Quiet Corner had been stable until the beach fronting them was depleted following the building of a sea wall at Black Rock, to the north, in 1939. Reduction of the beach allowed waves from the west to generate stronger longshore drifting, so that the beach fronting the bluffs gradually dwindled. By the 1970s storm waves were undercutting these bluffs, and erosion was threatening to undermine this part of the coastal highway. A proposal to extend the sea wall to halt this cliffing was opposed by local residents, who argued that a nourished beach should be

Figure 109 Intensive use of a nourished beach at Chiba on the shores of Tokyo Bay, Japan

put in as a protective formation which would also restore scenic and recreational values. In response, a beach terrace 100 m long, 25 m wide, and 1 m above high spring tide level was formed by pumping coarse shelly sand in from the sea floor during the winter of 1984 (Fig. 110).

Sandringham Beach, on the coast between Picnic Point and Red Bluff (Fig. 111), also stands in front of vegetated bluffs. In recent years the beach that fronted these has gradually diminished, and in late summer and early winter the southern part became so low and narrow that storm waves began to attack and undercut the vegetated bluffs, forming receding basal cliffs of slumping sandy clay. As these were cut back, there was a risk that a segment of the coastal highway, which here runs close to the top of the bluffs, would be undermined. In 1990 a beach 25 m wide and 600 m long was formed by trucking in about 35 000 m^3 of sand, placed to protect the base of the bluffs from further storm wave erosion. As a result the undercutting of the bluffs has ceased (see Fig. 128, p. 237). It should be emphasised that protection of a cliff base by means of a nourished beach will be effective as long as that beach is maintained, if necessary, by periodic renourishment.

These projects have demonstrated the importance of nourishing and maintaining a wide, high and persistent beach to prevent cliff recession. Such a beach is a means of absorbing wave energy and protecting the coastline from further erosion. It is important that a sufficient volume of

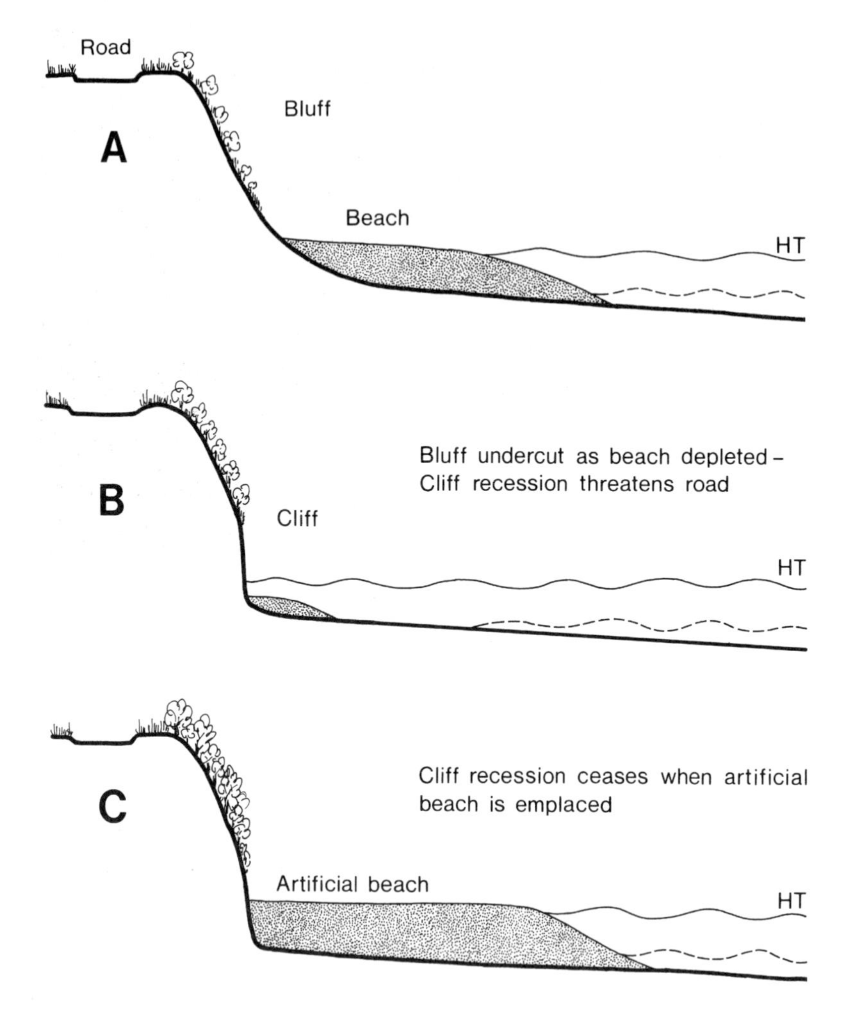

Figure 110 Changes on the coast south of Quiet Corner, Port Phillip Bay, Australia, where a previously stable bluff was undercut by marine erosion as the bordering beach was depleted, and this erosion was halted by the insertion of a beach to protect the cliff

beach material be maintained to protect the backshore, because the small quantity of sand or shingle that can be mobilised by storm waves can actually accelerate abrasion of cliffs or solid structures. This was the cause of severe erosion when the shingle beach was depleted at Hallsands in south-west England (p. 94). Increased abrasion occurred on the sea wall at Aberystwyth in Wales after much of the material used in a beach nourishment project quickly weathered and dispersed, leaving small quantities

Figure 111 The nourished beach at Sandringham, Port Phillip Bay, Australia, emplaced to halt the erosion of cliffs

of hard granitic gravel which were hurled at the wall by storm waves (So 1974).

Nourished beaches can also be used instead of solid structures to halt erosion on low-lying sectors of coastline (Fig. 112). At Cape Pitsunda, on the Georgian Black Sea coast, an eroding cuspate foreland consisting of sand and gravel beach ridges has been maintained by nourishing the beach.

Environmental impacts of beach nourishment

Beach nourishment requires the extraction of suitable fill material from a source area, its transportation to the shore and its deposition on a beach. Each of these procedures may have environmental impacts, because of sediment disturbance leading to increased turbidity in nearshore waters, benthic habitat disruption, the displacement or burial of plant and animal communities, and associated changes in oxygen, temperature, salinity, light, and the circulation of nutrients and chemicals in the sea and on the sea floor. Marine plant and animal communities may be reduced or destroyed, and their revival may be very slow, depending on the degree of disruption and the availability of recolonising biota (Pullen and Naqui 1983). Some of these effects can be reduced by planning the timing of dredging, transportation and delivery of sediment to nourish beaches. This depends partly on favourable weather conditions, and should also avoid breeding

Figure 112 Erosion of a low-lying marshy coastline near Gagra on the Georgian Black Sea coast (above) was halted by the use of beach nourishment (below)

seasons so that the impact on benthic biota is less severe. Reference has been made to the proliferation of nearshore seagrass as a consequence of bulldozing sand up on to the beach at Rosebud, Victoria, Australia (p. 185).

Source impacts

Dredging of sediment from the sea floor is widely used as a source of sand and gravel for building, road-making and other constructional work as well as material for beach nourishment. Such dredging disrupts benthic ecosystems, in particular submarine vegetation such as seagrass beds. It is necessary to select areas for sea floor dredging that are well away from critical habitats, breeding and feeding areas. Biological surveys should be made before dredging begins, and sensitive areas such as coral reefs, seagrass areas, and habitats for fish and shellfish should be mapped.

Extraction of sand from the nearshore zone to nourish a recreational beach in front of the Promenade de la Plage, at Prado, near Marseille in southern France, between 1974 and 1982 led to destruction of nearshore *Posidonia* beds (Rouch and Bellessort 1990). Eroding beaches on the shores of Hel Spit on the coast of Poland were nourished with sand dredged from the floor of Puck Bay, a lagoon to the south, and pumped across the spit, but this was stopped when it became clear that the dredging was damaging vegetation, increasing turbidity, and reducing fish populations in the lagoon ecosystem. Sand was then obtained from deposits on the floor of the Baltic Sea to the north and pumped in to the shore (Basinski 1994).

It is necessary to ensure that sea floor excavations do not excessively deepen nearshore areas, because this can lead to increases in wave energy, initiating or accelerating coastal erosion. Deep excavations also become stagnant hollows which are anaerobic and ecologically unproductive. Shallow dredging over a larger area may be initially more damaging, but ecological recovery is much quicker on shallow excavations. This was illustrated on the German North Sea coast, where beach nourishment on the island of Sylt began in 1972, using sand excavated from the nearshore sea floor by a hopper dredge, which cuts to a depth of about 2 m and disturbs large areas of the sea floor. Deeper dredging, from an anchored hopper, could restrict disturbance to a much smaller area by cutting to a depth of 40 m, but it was decided that such deep excavations may have more severe adverse impacts on wave processes and marine ecosystems (Dette 1990).

When sea floor sources of sand were sought for the nourishment of Mentone Beach, Port Phillip Bay, Australia, in 1976, there were fears that the dredging of sand would have adverse effects on benthic ecosystems, notably seagrass communities, a habitat for fish and shellfish. However, there was reassurance from Watson (1973), who had found that excavation of a trench 5 m wide and up to 3 m deep to carry a gas pipeline across the floor of Port Phillip Bay in 1972 (see Fig. 81) had caused only temporary depletion of benthic organisms, and may even have enriched the local fishery. In the event, the replenishment of Mentone beach does not appear to have adversely affected the sea floor ecosystems in Port Phillip Bay, although corrosion of the pipeline may yet pose a problem.

Dredging of sediment from the sea floor may release toxic materials. At Bogue Banks, North Carolina, material dredged from a harbour included fine sediment laden with hydrogen sulphide, which caused much intertidal and nearshore turbidity, modifying the habitat and killing many benthic macroinvertebrates. These disappeared during the 6 months of dredge disposal, and began to recover only slowly after beach dumping ended (Reilly and Bellis 1983).

Monitoring of sea floor plant and animal communities has shown that many gradually recover after dredging has ceased. In Florida, surveys showed good recovery of benthic biota 5 years after the dredging of sea floor areas off Hillsboro Beach (Marsh and Turbeville 1981).

Apart from dredging, benthic ecosystems may be damaged by burial or turbidity when sediment dredged from harbours or harbour approaches is dumped out on the sea floor. It may be better to use such sediment for beach nourishment or land reclamation rather than dump it offshore, where it can have ecologically adverse impacts on the vegetation that sustains the benthic fauna, including fisheries.

Impacts during transportation

Sediment dredged from the sea floor has to be transported to the shore, either in pipes or on boats. Leakages from pipes during pumping or losses as boats are loaded, navigated and unloaded can cause turbidity in the sea, the coarser gravel and sand settling quickly but the fine-grained silt and clay remaining in suspension. These effects can diminish light penetration and so disadvantage sea floor vegetation, while blanketing by spilt sediment has damaged seagrass communities, coral reefs, and fish and shellfish resources.

Some of the early beach nourishment projects in the United States used sediment obtained from dredging nearby bays and lagoons. These contained high proportions of fine-grained sediment, the release of which buried, or proved damaging to, benthic biota (Reilly and Bellis 1983). Later use of coarser sediment from offshore caused less damage to estuarine and nearshore ecosystems (Marsh and Turbeville 1981, Lankford and Baca 1989).

Mention has been made of problems of overland transportation, particularly lorry traffic passing through seaside resorts (p. 181).

Impacts of beach emplacement

Nourishment of beaches can have ecological impacts. In some places erosion of the preceding beach laid bare rocky or muddy areas and extended the habitat for the organisms that inhabit these. Such communities are disadvantaged when the beach is restored.

Beaches, particularly sandy beaches, are habitats for various crustaceans, worms, insects and birds, and are ecosystems adapted to such natural beach changes as cut and fill and longshore drifting, as well as tidal oscillations and frequent variations in wave energy and turbidity. The management of beach ecosystems is discussed below (p. 229). They are modified as beaches are eroded because the habitat diminishes, and further changes (perhaps revival) will occur as nourishment proceeds. Little is known of the impacts of beach nourishment on the various species, and this can only be determined by biological monitoring. Studies of ecological changes on nourished beaches require monitoring of organisms along transects across beaches before and after their nourishment (Nelson 1993). At Myrtle Beach in South Carolina, sand quarried from inland was trucked to the shore in early spring, and its deposition caused initial reductions in beach organisms, but there was then rapid recovery and after 4 months some sites actually showed species enrichment (Baca and Lankford 1988).

There is concern in Florida that nourished beaches may become too compacted for such activities as the nesting of Atlantic turtles (p. 230). Increased accretion downdrift from nourished beaches may have adverse effects on ecosystems, as at Saintes Maries de la Mer, on the southern coast of France, where sand is moving on to the salt marshes of the Camargue. At Rapid Bay in South Australia sand and gravel eroded from a beach formed by quarry waste dumping (p. 15) has drifted northward across the sea floor, blanketing formerly rich ecosystems on reefs and impoverishing the local fishery (Bourman 1990).

Artificial coastlines have become extensive in Singapore, where land reclamation has increased the area of the island by about 10%. On the northern (Strait of Johore) coast at Changi some nearshore areas were reclaimed by dumping earth and weathered rock, and their seaward edges were left unprotected so that wave action could sort the material and form beaches. Longshore drifting carried some of the reworked sediment to downcoast sectors, where beaches were improved if the sediment received was sandy, but spoiled where they were blanketed with silt and clay (Wong 1985).

The effects of sediment drifting along the shore from a nourished beach may also be seen as environmentally adverse. On the coast of South Australia, breakwaters built to protect the harbour at Port Macdonnell caused accretion of sand drifting from the west, infilling of the harbour, and erosion of beaches to the east. When it was suggested that the eastern beach be restored by nourishment with sand dredged from the harbour, local people objected that the dumped sand would fill nearshore hollows that they use for bathing and fishing.

Assessment of beach performance

Changes on nourished beaches are often rapid, and there is inevitably disappointment and criticism when an emplaced beach quickly diminishes. There has been much discussion of beach nourishment performance, particularly on the coasts of the United States. Before the 1950s beach nourishment projects on the Atlantic coast were intuitive, without much planning or design, and there seemed to be an assumption that the sandy beaches between New York and Miami were all more or less the same. Subsequently more attention was given to scientific research, acknowledging that there are variations in beach morphology, aspect and nearshore conditions, as well as contrasts related to the location and dynamics of tidal inlets. Most Atlantic coast beaches are of sand washed in from the sea floor, but there are some fluvially fed sectors and in the north some areas of cliff-derived beach sediment.

Engineering techniques have improved, and there has been increasing use of computer-based modelling since 1970 (Chou et al. 1983). Nevertheless, there are still doubts about the durability of beach nourishment projects. Walton and Purpura (1977) found that several nourished beaches on the Atlantic coast had performed poorly, and this they attributed to the widespread use of undersized material, nourishment too close to tidal inlets, and unexpectedly frequent storm activity. Pilkey and Clayton (1987, 1989) critically reviewed more than 90 beach nourishment projects on the Atlantic coast, and found that most of them had proved far more costly than anticipated and few had persisted as long as originally predicted. South of Cape Kennedy, engineers had been more successful in predicting the fate of replenished beaches, with Miami Beach a notable success, but on most of the nourished beaches on the Atlantic coast the sand deposited had been completely washed away in less than 5 years (26% in less than 1 year), usually because of erosion during storms; only 12% had persisted for more than 5 years (Leonard et al. 1990a,b). Moreover, the nourished beaches had not recovered from storms as well as natural beaches.

In an editorial in the *Journal of Coastal Research* in 1990, Pilkey noted that storms seemed to have been the major factor determining nourished beach longevity on the Atlantic coast, unpredicted erosion often being attributed to unusual storm activity. The public had been told that a replenished beach would recover during fair weather, that loss rates would diminish over time, and that the lost sand had moved offshore and would diminish wave energy on the depleted beach, so that the next beach nourishment would last longer.

Questioning the success of Atlantic coast beach nourishment projects le to a spirited discussion in the *Journal of Coastal Research* by Houston (1990), Pilkey and Leonard (1990) and Houston (1991a, b). This indicated that documentation of beach changes, both before and after nourishment, had been inadequate, that there was a need for the public to be better informed on how nourished beaches may perform, and a need to support accurate and sustained monitoring. A US Army Corps of Engineers (1994) report examined more than 100 replenished beaches and concluded that actual costs and volumes of sand placed were within 5% of predicted values (Houston 1995, Sudar et al. 1995), but Pilkey (1995) cited some omitted problems, including the fact that some beaches were severely depleted between nourishments. At Tybee Beach, Georgia, where the costs and volumes of beach nourishment in 1976 and 1987 were indeed less than predicted, the first placement disappeared within a year, so that for a decade Tybee was without any beach. Damage done to backshore property and structures should really be included as a cost item. Predictions remained uncertain: the 1993 nourishment of Folly Beach, South Carolina, was expected to require renourishment every 8 years, but it was already in need of replenishment after 1 year (Pilkey 1995).

It is now generally acknowledged that nourished beaches will be eroded, and will have to be replaced at intervals, and that this will require substantial and ongoing expenditure by governments and coastal communities. The alternatives are to revert to the use of solid protective structures, which do not co-exist well with beaches, or to allow natural changes to proceed on the coastline, abandoning eroding land. It seems likely that demands for beach nourishment will continue as a component of comprehensive coastal management programmes, because of increasing coastal population and development stimulating further demands for beach recreation, because of greater public awareness of beach erosion problems, and because of widespread opposition to the use of hard structures in coastal protection. Objections to nearshore dredging and truck traffic as means of obtaining and transporting sediment for beach nourishment, noted in Adelaide, South Australia, are likely to fade as demands for beach nourishment intensify. On the barrier islands of the Atlantic coast of the United States, local residents now regard truck traffic as acceptable if it is the only way of maintaining their beaches.

Public demands for beach nourishment are often a response to obvious depletion by storms, and the fear that further losses would rapidly ensue. At Long Island Beach, New Jersey, such demands led to nourishment of a depleted beach in 1979 (p. 191) but the added material disappeared after seven years. There is a divergence of longshore drifting here, sand from the northern part of the beach being washed back into Barnegat Inlet (from which it had been obtained) while sand from the southern part moved away southward along the coast (Ashley et al. 1987). Although narrow, this

tively stable, except for brief and temporary depletions
ods, and it is now locally regarded as acceptable. Public
ion hazards can be fickle: in the words of Pilkey and
1417), 'community apprehension over a narrow beach
and the absence of storms'. There was also the feeling
ld, if necessary, be restored again, this time with sand deposition concentrated in the zone of divergence, and losses northward and southward perhaps delayed by the insertion of a groyne field.

One of the requirements of beach nourishment is the use of sediment at least as coarse as that on pre-existing natural beaches, because finer material is quickly lost. Coarse beaches are less attractive to beach users (Campbell and Beachler 1984), but it would be simpler if they adapted to the use of more durable shingle beaches, of the kind already used for recreation at seaside resorts such as Brighton in southern England, Dieppe in northern France, and Nice on the French Riviera.

Many coastal countries are developing beach nourishment projects. In the United States, the Army Corps of Engineers is authorised to assist local coastal management agencies with beach nourishment projects that 'improve and protect publicly owned shores against erosion by waves and currents'. The need to integrate beach nourishment plans and practices in the broad framework of coastal management has been recognised by the United States Congress, which has increased financial incentives for beach nourishment (Davison 1992).

Costs and benefits

Beach nourishment is costly, but may be economically justifiable on sectors of the coastline, such as seaside resorts, where the emplaced beaches will be much used. There are costs in seeking sources of material suitable for nourishment, in extracting, transporting and emplacing it on the shore, and in subsequent maintenance (Christiansen 1977). Because of changing currency values it is difficult to make comparisons, but on the Atlantic coast of the United States the cost of beach nourishment and maintenance in 1995 was up to $US500 000 per mile per year. Much depends on the available sources of material for beach nourishment, and the distance across which suitable material must be conveyed to the nourishment site. At Bournemouth in England a 5 mile beach was nourished in 1974–75 with sand dredged from a site several miles offshore at a cost of well over £1 million, but in 1988–89 renourishment with sand supplied as a product of dredging the adjacent entrance to Poole Harbour cost only £130 000. The coincidence of a need for dredging with a nearby demand for beach nourishment thus resulted in substantial savings for Bournemouth. Allowance should also be made for preceding surveys and research: the project at Rockaway Beach, New York, in 1975, which included extensive preceding studies, cost $US14.3 million.

There have been various estimates of the relative costs of solid engineering structures and beach nourishment. The costs of beach nourishment are generally lower and more evenly spread over time than those incurred with the building of solid structures, which also require maintenance, especially after they are damaged by storms. Nourished beaches are more flexible than artificial structures because the beach profile can adapt to hydrodynamic variations, such as cut-and-fill sequences or storm events, without the damage caused to sea walls, groynes and other structures. In Port Phillip Bay the cost of beach nourishment does not exceed that of building and maintaining solid structures if the nourished beach survives for more than 7 years, while on the Georgian Black Sea coast beach nourishment at a cost of 51.9 million roubles was about half the cost of the previous unsuccessful coast protection with solid structures.

The chief benefits of beach nourishment are the provision of improved scenic and recreational values and additional coastline protection against the effects of storms and surges. There is a reduction in cliff erosion and storm damage to such structures as esplanades, roads and buildings on the coast. Unlike structures such as sea walls and groynes, a nourished beach protects one sector without inducing erosion downdrift, and some of the sediment deposited to nourish a beach may be carried by longshore drifting to downdrift sectors, augmenting their beaches and thus improving protection for adjacent developed coasts and their communities. Thus sediment moving alongshore is not really ‘lost’ if it benefits adjacent beaches and coasts. Examples of this have been noted on the coast east of Bournemouth, and in Singapore. Losses from gravel beaches emplaced on the shores of Lake Michigan are mainly alongshore rather than offshore, with the benefit that they may widen beaches that protect sectors downdrift along the coast (Roellig 1989).

Beach nourishment improves the recreational resource by increasing the beach area. The widened beach is attractive to visitors because it is a more natural and more pleasant environment for recreation than a coastline dominated by sea walls, tetrapods, breakwaters and groynes. A successful beach nourishment project guarantees a seaside resort its beach, provides a more attractive tourist lure, and results in more visitors and increased income, compared with resorts that are losing their beaches, or have become excessively adorned with artificial structures (Dean 1987b).

When beach nourishment is under consideration there is a need to consider marine pollution problems. In north-western France, the beaches fringing the city of Brest were much reduced over the past century by dock and railway construction, land reclamation for industry, and the building of marinas. In 1978, 1 km of beach was emplaced at Moulin Blanc, using 67 000 m^3 of calcareous sand dredged from shoals offshore. The beach was built 100–120 m wide and 0.5–1.0 m above the high spring tide level. Up to 1700 people use this beach on a summer day, but unfortunately the nearshore waters are polluted by sewage and chemical wastes from nearby industry (Hallégouet and Guilcher 1990). The benefits of a nourished beach will clearly be greater if nourishment is preceded by action to ensure satisfactory nearshore water quality.

There is also the question of who should pay for beach nourishment projects. In most countries the cost is met by national or local government agencies on the grounds that the restored beaches are public facilities, but where the beaches are private, or public access is impeded for one reason or another, this becomes difficult to justify. In the 1970s beach erosion at Miami led to demands for public help, but by then 95% of the beach had passed into private ownership. The US Army Corps of Engineers then made unimpeded public access a condition of beach nourishment at public expense.

It has been found that nourishment of beaches on the east coast of the United States has been of benefit to property owners because of more effective protection from storm damage and the provision of an improved recreational resource (Olsen 1982). Prices of land and housing rise, as do incomes from rent and tourist expenditure, behind beaches widened by nourishment in South Carolina (Pompe and Rinehart 1994). On the Atlantic coast generally beach nourishment can increase real estate values by up to 21% (Black et al. 1988), and induce further development or rehabilitation of existing development (Stronge 1990, Bodge 1991).

Nevertheless, many coastal engineers remain cautious about the economic viability of beach nourishment projects. In the words of the British engineer M.G. Barrett (1989): 'I have no doubt that there is a consensus amongst coastal engineers that the ideal form of coastal defence in purely engineering terms is a massive beach. Whether this can be achieved in future economic terms and in proper long-term use of available resources are quite another matter'.

Effects of a rising sea level on nourished beaches

Predictions of global warming and a world-wide sea level rise now need to be taken into account in long-term planning for coastal management, including beach nourishment (Bird 1993b). Studies of the effects of a rising sea level on beach-fringed coasts have shown that erosion will be initiated or intensified as submergence proceeds (p. 79), except where there is a continuing natural or artificial supply of sediment to maintain beaches at progressively higher levels. The coastline can be maintained by building sea walls and other protective structures to prevent erosion and submergence, but as has been noted (p. 103), these are likely to cause further beach erosion, and in due course beaches (including those that have been nourished) will disappear, leaving the coastline artificial. It will be possible to maintain beaches by continuing renourishment as sea level rises, the limiting factors being the availability and cost of suitable nourishment material and the extent of hinterland submergence, which may have to be prevented by building sea walls to protect low-lying (polder) land behind the renourished beach, or building up coastal lowland levels by landfill to keep pace with the rising sea.

The forecast world-wide sea level rise is likely eventually to come to an end with a stillstand of the sea at a higher level, but predictions of what this level will be are not yet forthcoming. If they were, it would be possible to deduce the nature of the coastline that will form at the raised sea level, the possible extent of future beaches, and sectors where beaches could be nourished in advance. This principle was illustrated in Tasmania, where a dam was built to create a reservoir in a valley, submerging the former Lake Pedder. Before the water level rose, coarse sand was deposited on a sector of the contour (dam overspill level) it would reach, and this became a beach on the western shore of the enlarged Lake Pedder.

In South Australia the Coast Protection Board has considered whether the Adelaide beaches could be maintained if sea level rises, as forecasted, and found that beach nourishment would continue to be feasible, especially if inland sources of sand were used, until the sea rises 20–30 cm above its present level, but thereafter it may be necessary to construct major sea walls and accept that the beaches will disappear (Wynne 1984). Inevitably, there will be future cost increases of beach nourishment projects, with more frequent and more substantial filling. As Weggel (1986) remarked, 'if projections of an increasing rate of sea level rise are correct, it will become increasingly difficult to economically justify future beach nourishment projects'. If the sea level rises, beaches will diminish as the result of submergence and erosion, and by the year 2100 the only seaside resorts that still have beaches will be those where they have been artificially nourished. Those that are able to retain their beaches will continue to operate as tourist and visitor attractions; those that do not will have to cultivate other attractions if they are to survive.

Chapter Six

Beach use and management

AIMS OF BEACH MANAGEMENT

Beach management seeks to maintain or improve a beach as a recreational resource and a means of coast protection, while providing facilities that meet the needs and aspirations of those who use the beach. It includes the framing and policing of any necessary regulations, and decisions on the design and location of any structures needed to facilitate the use and enjoyment of the beach environment.

Previous chapters have presented background on beach processes, erosion problems, beach protection and beach nourishment. Beach managers should be able, with this background, to examine such questions as where the beach sediment came from, whether it is still coming, and whether the beach is stable, or shows long-term accretion or erosion, as distinct from short-term cut and fill or seasonal alternations. They should consider whether beach sediment is being gained or lost alongshore, coming in from the sea floor or disappearing seaward, or being gained from, or lost to, the backshore and hinterland. They should be able to assess the possible effects of stream outflows and tidal inlets on beach stability. If there are artificial structures, such as sea walls, groynes, breakwaters, boat ramps and drains, they should be able to discuss the effects these have had on the beach system, and what would happen if they were either removed or elaborated. They should have some awareness of whether beach nourishment is required, and if so what sources of beach material are available, what mode of delivery will be practical and economic, and what arrangements should be made for mapping and monitoring before, and particularly after, beach nourishment takes place.

With this background, attention can now be given to problems that arise from the use of beaches by people for various purposes.

JURISDICTION

Most of the world's beaches are regarded as public areas, although some are privately owned, particularly in the United States. In many countries the backshore has been retained as public land (Crown Land in British terminology), except in areas where it has been alienated for urban, suburban, commercial or industrial development. The landward boundary of the area for beach management may or may not include parts of the backshore, and where a beach is being reduced by erosion or increased by accretion there will be accompanying changes in such boundaries. Thus if property fences are maintained on a receding coast, part or all of any public beach frontage can eventually be claimed as private land unless arrangements have been made to set boundaries back in order to perpetuate a public reserve. On prograding beaches there must be a legal justification for taking over the new backshore land. Management of accreting beaches raises problems that differ from the management of stable or eroding beaches because of the provision of such new backshore terrain.

The seaward boundary of the area for beach management is conditioned by the lines of high and low tide, but many management agencies include part of the nearshore sea area, usually defined as a conventional distance seaward, because of their concern with features, activities and problems in shallow water adjacent to the beach, and with land–sea interactions. In most countries management agencies are coastal administrative divisions (such as counties, municipalities or communes), with boundaries that may or may not coincide with natural features (such as river mouths or prominent headlands) when they cross the coastline. Often such boundaries run across beach compartments, the management of which then requires co-operation between neighbouring administrations.

PEOPLE AND BEACHES

There are many kinds of beaches, and many kinds of people visit them in pursuit of a wide variety of activities. Some beaches have waves suitable for surfing; on others the sea is calmer and can be used for swimming. Some fringe stormy seas or water with powerful currents; others are shallow and safe for small children. Beaches are venues for picnics in many countries: for clambakes in New England, lifesaver parades, as in Sydney, Australia, and festivals, as in Penang, Malaysia. Some have firm sand suitable for beach games; others are steep and gravelly. Some have a high incidence of litter, pollution and seaweed; others stay clean, swept by waves and wind. Some have a wide range of facilities, such as car parks, toilets and showers, refreshment kiosks, lifesaving clubs, first aid centres, garbage bins, boat sheds and beach huts; others have a natural setting, often with a rich vegetative background. Some are close to shops, hotels, apartments or camping or caravanning grounds; others are remote and empty. Some have nearby vegetation or artificial shade and shelter from sun or sudden adverse weather; others are exposed to the various elements. Some have the fun of a crowd, like Blackpool or Coney Island, with shops and amusement arcades close by; others offer the pleasures of solitude in a natural setting. Beaches are often used to launch boats or haul them ashore, or to harvest washed-up seaweed. Patterns of visitor activity vary with the season, weather and tide conditions, weekdays, or weekends and public holidays. A major factor is accessibility. Beaches at or close to seaside towns or with highways feeding coastal car parks can become very crowded on a summer's day, and it may be necessary to drive and then walk a considerable distance to find a really isolated beach (Hecock 1970, Spaulsing 1973). Alternatively, one can escape all the problems of seaside beaches by going to the Japanese indoor beach stadium at Wild Blue, Yokohama, which generates waves that break on a rubberised sand-grained shelf under artificial sunlight at a steady temperature of 30°C. The entry fee is $US50, and another $US50 hires a beach chair and a surfboard.

BEACH ACCESS

Access to privately owned beaches, extensive on the coasts of the United States, requires permission from the owners, and may not be allowed (Fig. 113), particularly if the beach is backed by houses or enclosed areas with fences or walls down to the low tide line (Fig. 114). Where the beaches are owned by a commercial organisation an entry fee of several dollars may be charged. Fortunately many of the world's beaches follow the British and European tradition whereby public access is free, at least to the zone between high and low tide lines, which in Britain is mainly Crown Land. Access has nevertheless to be gained from the backshore, which may be in private ownership, with fenced or walled houses and other buildings, and may be confined to particular pathways with 'rights of way'. In Australia the backshore zone is generally a Crown Land Reserve* although some sectors have been alienated to private ownership, or allocated for specific exclusive uses such as clubhouses for various beach or sea activity groups, notably boat clubs. Management of Australian Coastal Reserves is integral to beach management, and is the responsibility of coastal municipalities or committees of management appointed to particular sectors, and one of the requirements is that public access to beaches be maintained, and indeed facilitated (Bird 1988). Other countries have variations on these

* Often known in Australia as Foreshore Reserves, a confusing legal term because the scientific definition of foreshore is the zone between high and low tide lines: the term Coastal Reserve is now widely preferred.

Figure 113 Private beach in Washington State, USA

Figure 114 Access to the beach can be impeded if private development occupies the backshore, as at Sorolmar on the Californian coast north of Los Angeles

Figure 115 Ramp designed to give wheelchair access to Beaumaris Beach, Port Phillip Bay, Australia

arrangements, most needing notices and fencing to indicate areas available to the public, including beach access routes. Happily the racial segregation formerly enforced in South Africa disappeared with the end of the apartheid policy.

Many beaches are backed by sea walls and esplanades, with access from roads and car parks by way of steps or ramps, which may include features designed to be used by disabled people (Fig. 115). On some sectors access is much restricted by backshore structures, as at Cleveleys, north of Blackpool, where a high bank with a capping fence obstructs not only the sea view, but also people wishing to get to the shore. Beaches backed by cliffs may require stairways or lifts to enable people to reach them. If a beach has a backshore of vegetated dunes or bluffs, pathways are worn by people coming and going, and it is necessary to prevent damage to the vegetation and the onset of erosion. Pathways across vegetated dunes may grow into blowouts excavated by onshore winds, and runoff after heavy rain may cut gullies down the lines of pathways that descend steep slopes. Fragile vegetation may need to be protected (fenced out) from public access, and walkways may have to be built to provide a means of reaching the beach. Grassy, shrubby or wooded dunes behind a beach also attract people looking for shelter from the wind, or shade from the sun, but with intensive use this vegetation cover may be destroyed and the dunes eroded. Problems of vehicle access through dunes to beaches are discussed below

Figure 116 Beach huts at Brighton, Port Phillip Bay, Australia

(p. 241). In practice, dune management is often linked with beach management, and further information can be obtained from dune management manuals, for example Ranwell and Boar (1986), Cullen and Bird (1980) or Doody (1985).

Various waterborne craft, such as canoes, surfskis, jet skis and rafts, may have to be taken across beaches to gain access to the sea, and larger sailing and power boats require a launching ramp or the use of a winch. Boats are often drawn up and parked on beaches where harbours are not locally available. Access to many beaches is impeded by coast roads and railways, and may require bridges, tunnels or pedestrian crossings with appropriate warnings and controls such as traffic lights. Beach huts (Fig. 116) can also impede access.

BEACH STRUCTURES

Apart from sea walls, groynes, offshore and undersea walls designed to keep a beach in place, as discussed in Chapter 4, various other structures are developed on or behind beaches that become much used for recreation. These include car parks (Fig. 117), roads, boat launching ramps, piers, toilets and shower facilities, refreshment kiosks, picnic tables and barbecues, lifesaving clubs, first aid centres, boatsheds, beach huts, fishermen's shacks and garbage bins. Often there are various other buildings, as well as

Figure 117 A car park built on to the upper part of the beach near Rosebud, Port Phillip Bay, Australia

pipes, drains, and power and phone lines. Beach management agencies usually require permits for excavation, or construction of solid structures, which impose conditions concerning their impacts on vegetation and the risks of erosion due to interactions with wind and wave action or runoff. There is a preference for structures that are expendable or removable if erosion threatens them, such as beach huts, shelter sheds and life-saving observation towers. Many beaches are backed by urban or suburban seaside development, with parks and gardens, caravan and camping grounds, golf courses, sports facilities and other resort buildings.

There are social and cultural variations in the acceptability of certain beach structures. On Mediterranean beaches, particularly in Italy, there is considerable beach furniture, including cafés, huts, walkways, tables and fixed umbrellas, supporting intensive and highly commercialised beach use, at least in the summer season. In northern and western Europe, and most other parts of the world, this kind of development is regarded as unnecessary clutter, spoiling the beach environment and over-organising beach recreation. Structures related to boat use are often more acceptable. These include boat sheds, mechanical devices such as winches and chains used to haul boats ashore, as on the shingle beaches at Deal and Hastings in south-east England, and structures used in drying and repairing fishing nets. In Scandinavia there are backshore huts used for the drying, curing and smoking of fish.

Figure 118 Anglers at Bexington, Chesil Beach, on the south coast of England

BEACH ACTIVITIES

Surveys of beach users show that most people come to the beach to enjoy themselves, relax or exercise. Principal activities are swimming, surfing, snorkelling, paddling and playing in the water (often with floats or rafts), or windsurfing (officially known as boardsailing), surfskiing, sailing, canoeing, motor boating, waterskiing or jetboating in the nearby sea. Some try sea angling (Fig. 118), others search the shallows with a net for shrimps, or hunt shellfish. Children play on the beach with bucket and spade, making sandcastles or digging canals. People sunbake, or go walking, running, horse-riding, or flying kites or model aircraft. Other activities include cycling, sightseeing, picnicking, sitting talking or reading, meeting people, looking at the opposite sex, being amorous and falling asleep. Deck chairs, beach umbrellas and wind screens are the commonest portable furniture. Beach games include throwing and catching a ball, frisbee or boomerang, or playing cricket, football, rounders or volleyball. Beach volleyball, propelling a lightweight· ball to and fro across a net, is particularly appropriate for a beach setting because soft sand permits more spectacular retrievals than on a hard court. Racehorses are exercised on Mordialloc beach in Victoria, Australia.

Firmer sand, particularly where there is a wide beach at low tide, offers opportunities for sand (or land) yachting, in which 3- or 4-wheeled vehicles with sails are propelled by the wind. Sailing of such vehicles originated many centuries ago on land in China and Egypt, and became popular on wide North Sea coast beaches in the Netherlands, Germany and Denmark in the 16th century. Races are held on beaches where the sand exposed at low tide is hard and firm, as on the Lancashire coast at Lytham St. Anne's, where the first World Championships were held in 1970. The sport, which has an international federation in Brussels, has become popular on some North American beaches in recent years. The yachts can move at speeds of up to 100 km/h, and it is necessary to designate beach areas for their exclusive use in order to avoid accidents involving other beachgoers. There are similar problems where cars or motor cycles are driven on beaches.

Lifesaving clubs arrange drilling and simulate rescues on and from beaches, and there are surfing and other carnivals and competitions, as well as traditional beach entertainments such as Punch & Judy, donkey rides and sand castle competitions. Vendors of food, drink and ice cream may move on to the beach or backshore, and some like to have vendors' vehicles bring refreshments on to the beach.

The atmosphere of a European beach vacation was memorably captured in Jacques Tati's film, *Monsieur Hulot's Holiday*. Among many works that express the beach experience are Ruth Manning-Saunders' *Seaside England* (1951), Sarah Howell's *The Seaside* (1974) and David Gentleman's *Coastline* (1988), while in Australia, which prizes its beaches, Geoffrey Dutton's *Sea, Surf and Sand: The Myth of the Beach* (1985), Robert Drewe's *The Bodysurfers* (1987) and Philip Drew's *The Coast Dwellers* (1994) variously explore aspects and implications of the beach culture, including art, literature and films.

Just as there are many different kinds of beach, so there are many different attitudes towards the beach environment and how it should be used. Some people seek a crowd, others seclusion. A survey of beach use on the Spanish coast showed that the beaches were seen principally as places for passive activity (sunbathing, doing nothing, watching others), an escape from everyday activities. Beaches were expected to be clean, comfortable, friendly and safe places, with life-saving, first-aid and police. Some wanted the beaches left natural, with facilities having minimum impact on backshore landscapes; others (usually younger and less well educated) wanted more recreational structures, such as water slides, bars and nightclubs. Some wanted sufficient isolation and privacy for nudity. Many complained about traffic congestion, dogs and noisy radios. Some wanted to forbid beach games and motor boats near swimming areas, but 15% wanted no restrictions of any kind, believing that the beach should be a place of freedom (Fabbri 1990).

Beaches are a major open-air recreation resource for most coastal towns, although they may be little used in winter. They are regarded as healthy

Figure 119 Strandkorben on the beach at Sylt, Germany

places (Skegness, as a famous poster told, is 'so bracing') and as open space – a rare resource in many countries. They often provide better opportunities for walking along the shore than inland, where substantial public open spaces and rights of way are sparse. If beaches are less than about a kilometre in length people tend to walk towards each end, perhaps a rocky promontory or breakwater, but longer beaches may discourage walkers.

There are marked cultural contrasts in beach use around the world. Primitive people, especially in the tropics, often use the beach and the sea as a toilet and washing place, while visitors to Bali beaches are sometimes disconcerted to find a funeral in progress in the sea. In India and south-east Asia urban beaches become busy and crowded, with many hawkers and gypsies. On tropical beaches in Tanzania and Ecuador fishermen clean and sell their catch. Northern and western Europeans seem to become more relaxed and friendly when they venture on to their beaches, but on the German North Sea coast there is a tradition of each family or group building a *sandburg*, a circular wall of sand in which they can place their *strandkorb* (beach basket seat) as a shelter from sun, wind and rain (Fig. 119). The sand rampart is maintained by sprinkling it with sea water to prevent it drying and crumbling, and sometimes it is adorned with sculptured mermaids, anchors, battlements or messages. There is an implication of territorial rights, and vigorous arguments can take place about the ownership of these structures when the next day brings new visitors.

It seems likely that when most people think of a beach they imagine a stretch of sand beside a blue sea, backed by waving palm trees, the kind of scene portrayed in tourist literature. Yet there is a world-wide trend for backshore development, and natural beaches are becoming very scarce – even in coastal reserves and National Parks there is a tendency for people to expect car parks and at least some recreation facilities. As world population grows (and coastal populations are growing more quickly than the average), urbanisation and the demand for backshore development is causing more and more beaches to become fringes of developed land. Already, completely undeveloped beaches are difficult to find on the coasts of warm countries or those with a season warm enough for substantial beach use.

Surveys of beachgoers often list 'watching other people' as a beach activity. The beach is an environment where it is more acceptable to take off most (in places all) of your clothes than in most inland public spaces. In consequence, it is a place where, usually free of charge, one can watch and admire (or envy) beautiful and healthy people. There are contrasts in beach-using populations from country to country – certainly in some countries the beach-using population includes many overweight and unfit people, whereas in others people in this category may shun the beach, fearful of ridicule or scorn. Where this is the case (probably in Australia, for example), the existence of a beach culture may actually have positive feedback in terms of general levels of health and fitness. However, recent awareness that exposure to strong sunlight is a health hazard, causing sunburn, skin damage and skin cancer, may have diminished beach use (even in Australia), and could lead to increasing demands for beach and backshore shade, either from planted trees or artificial shelters.

Surfing

Surfing requires particular wave conditions, and is best where ocean swell moves in towards a beach. It has been noted (p. 28) that south-westerly swell from the Southern Ocean generates surf on the coasts of South Africa, Australia, New Zealand and South America and is transmitted across the Pacific Ocean to western North America and across the Indian Ocean to Thailand and the southern shores of Indonesia. Other oceanic coasts receive swell from various directions, as in Hawaii.

Surfers prefer beaches where the waves are steep, smooth and high, and about to break (Fig. 120). Suitable waves arrive on coasts with a long

Figure 120 (*opposite*) Diagrams indicating favourable wave conditions for surfing. (A) When gently refracted waves on a gently shelving shore break parallel to the beach; (B) when waves are raised over a submarine reef or shoal; (C) when waves are diffracted through a gap in a nearshore reef; (D) when waves arriving obliquely steepen and break laterally as they move towards the beach

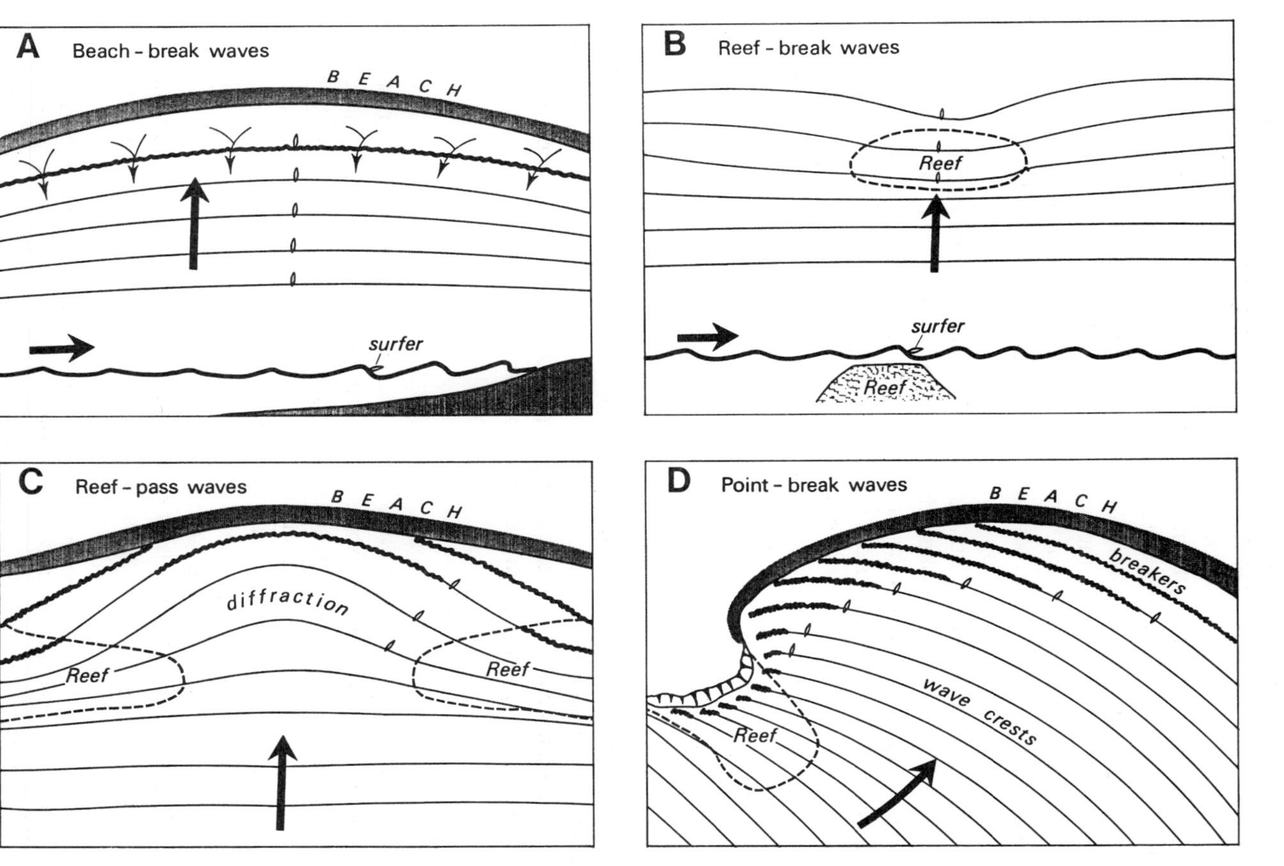

Figure 120 *For caption see opposite*

oceanic fetch, where the nearshore water is relatively deep and the sea floor profile rises steadily shorewards. In some places nearshore reefs or rocky outcrops may enlarge and steepen incoming waves. Under these conditions ocean swell may form waves several metres high, and surfboard riders aim to catch the shoreward current in the upper wave front. The most consistent waves are those that come in over a hard rocky sea floor, a sandy floor being apt to vary considerably as sand bars form, migrate, change shape and disperse, so that the waves are more variable. The most valued surfing waves are those that arrive obliquely to the shore and break laterally, so that the surfer can edge along the wave front or even pass through tunnels in the spilling surf.

Surfing has little impact on beaches, but where it is popular there are demands for backshore buildings for Surf Life-Saving Clubs, car parks and other supporting facilities such as kiosks, toilet blocks and first-aid centres, and eventually a surf beach coast may be converted into a kind of sports stadium. During surf competitions it may be necessary to fence off parts of beaches, and to charge fees for access and car parking (Bird 1993c).

BEACH PROBLEMS

Seaweed, shells and pebbles

Various natural materials are washed up on to beaches, particularly during and after storms. Shells and pumice are components of some beach sediments (pp. 13, 18), and the bones of fish, marine mammals and sea birds are often encountered, but usually the most abundant material is seaweed, including the plants known as seagrasses. Such vegetation grows on intertidal and subtidal rocky, sandy and muddy substrates, and is torn off during storms, particularly towards the end of the summer growing season, and then washed onshore. Many beaches are heaped with seaweed, ranging from the large kelp to smaller plants such as sea lettuce, wrack and the various seagrasses. The abundance of this natural litter may be increased when nearshore waters are enriched by nutrients as the result of the disposal of sewage in the sea, runoff containing fertilisers from farmland, and other organic wastes. Mention has been made of a case of increased seagrass litter resulting from accidental habitat improvement by nearshore bulldozing (p. 185).

Seaweed on beaches, or floating in nearshore waters, is a nuisance on recreational beaches: it is unsightly, smelly and infested by insects, especially flies; it can impede access to the water and obstruct swimmers and surfers. In some countries, notably Japan, seaweed is harvested from the shore as a food source, while on King Island, Tasmania, kelp is raked from the beaches for export to Scotland to be processed into alginates.

Beaches may be modified where seaweed is collected from them mechanically, partly by the impact pressures of tractors and lorries, and partly because sand and gravel may be included in the material taken away from the shore.

Some seaside resorts want their beaches to consist entirely of sand, and go to the trouble of removing pebbles and shells, as well as seaweed, just before the holiday season begins. At Weymouth, England, pebbles from a shingle beach to the east drift along the shore, and at the end of winter some of them are scattered on the sandy beach near the Queen Victoria Jubilee Clock Tower. They are then removed. Shelly material brought in with sand dredged from the sea floor has been collected and removed from nourished beaches on the north-east coast of Port Philip Bay, Australia.

Litter

In addition to natural materials, many beaches carry a remarkable variety of litter, notably plastic containers and bags, cans, glass, brick, wood and paper (Fig. 121). Near ports and coastal urban areas such alien material can form a significant component of beach sediments. Some of it has been dropped by beach visitors, but much is washed out from rivers and drains during runoff after heavy rain, and large quantities drift along the shore from ports and industrial areas, or from coastal garbage dumps or reclamation zones where waste material is used as landfill. Some litter is flotsam, washed onshore; some is jetsam – rope, packaging, cans and stoppered bottles dropped from ships and smaller boats – that floats in on the waves. After storms garbage is often washed up, mixed with seaweed, from the sea floor, and items washed in from fishing boats include craypots, fish boxes and crates, bait baskets and nets (Wace and Zann 1995). Beachcombing, the search for items of value along the shore, was worthwhile in earlier centuries, when much of it came from shipwrecks, but now there are few rewards among the trash. Perhaps a better return is obtained by using metal detectors to find coins and other objects lost on resort beaches.

Beaches on the east coast of Honshu, Japan, within 100 km north and south of Tokyo, are heavily littered, especially with wooden, plastic and paper packaging, and pollution of this kind is often encountered close to other major ports, or in the vicinity of coastal garbage dumps. Similar problems have been reported from Israel (Golik and Gertner 1992). Even on remote beaches, such as oceanic atolls, where locally generated litter is rare, it is possible to find a wide variety of waste items. Beaches on the north coast of the Chatham Islands, east of New Zealand in the South Pacific, carry large amounts of litter, much of it labelled in Chinese and emanating directly or indirectly from east Asia. Surveys of Anxious Bay, on the remote western coast of Eyre Peninsula, South Australia, by the South Australian Scientific Expeditions Group cleared a 26 km beach in

Figure 121 Litter and seaweed strewn on a beach after a storm, Sandringham, Port Phillip Bay, Australia

Anxious Bay in 3 days, collecting 344 kg of litter in 1991, 391 kg in 1992 and 216 kg in 1993, averaging 8–15 kg/km: about 60% was plastic and 30% glass, and some of the items came from South America and South Africa (Wace 1994). Apart from being unsightly, litter can be dangerous: broken glass, jagged cans and drug syringes are unwelcome items on many much used beaches.

Surveys on British beaches have found litter that has come in substantial quantities from North America, including items tipped over Canadian cliffs into the sea. Garbage from motor boats and yachts is more abundant than ship junk, the disposal of which is prohibited by the international Marpol Convention, signed by most of the world's shipping nations in 1988.

A substantial, clean, well-managed beach is regarded as essential for a seaside resort. The obvious response to the litter problem is regular beach

Figure 122 Beach cleaning in progress at Beaumaris, Port Phillip Bay

cleaning, either by collectors or using mechanical devices (Fig. 122). In Singapore there is daily cleaning of the sandy beaches that have been nourished for recreational use at East Coast Park, Changi, Pasir Ris and Sentosa (Wong 1991). Vehicles designed to clear litter and seaweed from a sandy beach inevitably remove some sand and also compact the beach, thereby contributing to erosion (p. 77). Much litter is washed away by waves and tides, only to be deposited elsewhere. Some countries have anti-litter campaigns, aiming to persuade beach users to deposit their waste in backshore or beach exit receptacles or take it home, but as has been noted that much beach litter comes from the sea, from drains, or along the shore. In Australia efforts have been made to interest people, particularly scouts, guides and schoolchildren, in collecting litter as a beach game.

Pollution

Apart from litter, a major cause of beach pollution is the discharge of oil waste from ships offshore, forming lumps of black or dark brown tar on the beach. These are a hazard for water birds, which can be immobilised and killed by oil. Tar also has an unpleasant smell, and can soil skin, footwear and clothing. Nearshore sea pollution by oil slicks is unpleasant for swimmers. The problem was severe on the beaches of the English Channel following the Torrey Canyon oil spill on Seven Stones near the Isles of Scilly in 1967, and the wrecking of the Amoco Cadiz oil tanker on the shores of Finistère in 1978. In the Caribbean, many beaches have tar deposits derived from Gulf of Mexico oil wells or spillage from tankers (Jones and Bacon 1980).

The arrival of waste from sewers opening on to the shore or into the sea can result in malodorous material on beaches or in the water, and under these conditions swimmers risk infections and sickness. Numerical standards have been set (for example by the European Union) for acceptable levels of pollution in sea water at coastal resorts. An indicator of sewage pollution is the presence in large numbers of the bacteria *Escherichia coli*, which multiplies in human and animal faeces. Conventionally concentrations of less than 100 *E. coli* per 100 ml of water are accepted as safe for bathing. Such measures have been used by a number of agencies, including the European Commission, to classify beaches in terms of bathing water quality, and resorts which fail to meet the standard set are under pressure to improve their sewage treatment and discharge systems. In 1994 nearly a quarter of the 400 registered swimming beaches in the United Kingdom failed to attain the standard set for concentrations of bacteria from sewage. Such failures have prompted water companies to spend large sums on new or improved treatment works, for example at Southport and Blackpool in Lancashire. Classifications of bathing water quality certainly affect tourism in beach resort areas, and have led to such organisations as Beachwatch, set up by the British Marine Conservation Society, and publications such as the annual Good Beach Guide (Linley-Adams 1994), which also locates the beaches where untreated sewage is released into the sea. There are also reports by the European Commission and the UK Consumers' Association on European bathing beaches. Each summer several resorts in Europe (notably around the Mediterranean Sea) receive adverse publicity because their beaches fail to reach satisfactory water quality, usually because of nearby discharge of raw sewage into the sea, but sometimes because of industrial pollution. While much attention has been given to sewage pollution, it should be noted that other organisms, including viruses, cause health problems in bathing water.

Other forms of beach pollution include toxic materials derived from industrial sources. The beach north of Bunbury in Western Australia was stained red by effluent from an industrial plant, and chemical effluents have discoloured the sea and rendered shores abiotic in many places, such as Burnie in Tasmania. In August 1987, New Jersey beaches were closed for several days following pollution by hospital waste, which had apparently been dumped in the sea off New York.

Algal blooms

Large quantities of dissolved chemicals (notably phosphates and nitrates), both natural leachates from the soil and weathered rock outcrops and substances produced by human activities, such as sewage, sullage, fertilisers and pesticides, flow into the sea. Water containing these chemicals is discharged by rivers or drains, and provides a source of nutrients for marine plant and animal life, including fisheries. An increase in nutrient status (known as eutrophication) may be initially beneficial, increasing the productivity of marine ecosystems, but excessive nutrients can lead to the development of algal blooms, a rapidly increasing and expanding algal population, especially in calm nearshore water.

Algal blooms include the so-called red tides that occur in the Mediterranean and other warm seas. Similar blooms of blue-green algae (e.g. *Nodularia* spp.) may occur in estuaries, lagoons and landlocked bays. They have formed in such places as Moriches Inlet on Long Island, New York, in Oslo fjord, in Lake Erie, and in coastal lagoons such as the Gippsland Lakes in Australia. Where algal blooms develop nearshore and spread along a beach the presence of cloudy, coloured water may deter swimmers. Those who do swim may acquire infections, particularly of the ear, nose or throat, and stomach upsets. The appearance of a severe algal bloom along a beach may justify temporary restrictions on swimming, with signs that indicate sectors of beach that are affected. Where beach pollution is a recurrent problem it may be necessary to identify sources (rivers, pipes, drains) and to put up signs to indicate 'black spots' and zones likely to be affected.

MANAGEMENT OF BEACH ECOSYSTEMS

It has long been acknowledged that beaches, especially sandy beaches, have a range of chemical and biological characteristics, in addition to their obvious physical features. In general, fine sand beaches are often richer in associated chemicals, notably organic carbon, nitrates and phosphates, than coarser beaches, especially if they receive accessions of seaweed and other organic products washed in from the sea or carried down from the hinterland. Various studies have shown that contrasts in the chemical characteristics of neighbouring beaches are accompanied by differences in biological, and particularly microbiological, associates, for example in Anglesey in North Wales (Pugh et al. 1974).

The importance of beaches as ecosystems has been recognised, notably as a sequel to an important conference in South Africa in 1983 (McLachlan and Erasmus 1983). Beaches provide habitats for a variety of plants and animals adapted to natural changes: the possible effects of beach nourishment were mentioned on p. 201. Sandy beaches are inhabited by burrowing organisms such as crabs, shellfish and worms, together with an often rich microbiota. The drier upper beach has air-breathing organisms, whereas the wetter intertidal zone has species that can survive frequent waterlogging.

Figure 123 Reeds (*Phragmites communis*) spreading on to a sandy beach at Hel Spit, on the Polish coast

The broad surf zones of dissipative beaches have richer ecosystems, nourished mainly by diatoms and nearshore algal blooms, than the narrower surf zones of reflective beaches (Brown and McLachlan 1990). On gravel beaches the upper, more stable areas may be colonised by lichens, mosses, grasses and shrubs. Invasion of beaches by reeds and rushes has been noted in recent decades on the shores of the Baltic, particularly on sectors where sandy beaches have been lowered by erosion to levels that are kept wet by seeping groundwater (Figs. 123 and 124) (Bird 1990b).

Many bird species are seen on beaches, as well as on nearby cliffs, rocky shores, lagoons and marshes. Where birds nest on the beach or backshore it may be necessary to place warning signs and fencing to prevent disturbance by visitors. During the spring nesting period public access is limited to the terneries on parts of the British coast, as at Blakeney Point, Norfolk, and areas of beach where birds are breeding are then fenced off from public access (Fig. 125). On some beaches netting has been used to deter birds from nesting in areas likely to be consumed by erosion during the nesting season.

Each season, turtles come ashore to lay their eggs within the beach at a number of places around the world, including sites in Mexico, Florida and the east coast of Malaysia. Tourists have been encouraged to watch them make their way up on to the beach and excavate hollows in which the eggs are laid. There is a risk of disturbance as this takes place, of eggs being taken by collectors or destroyed by trampling, and of the beach being

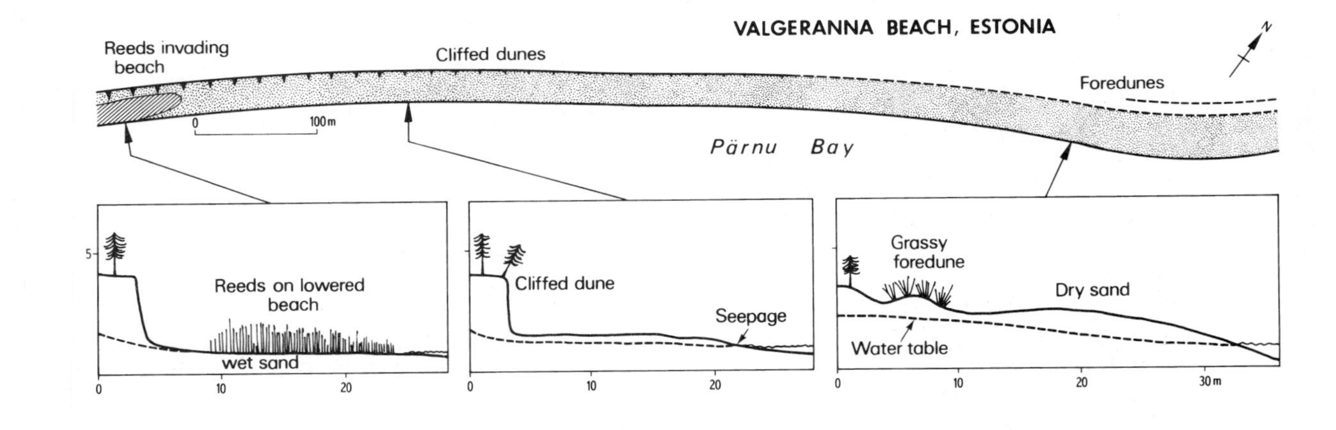

Figure 124 Profiles on Valgeranna Beach, Estonia, showing reed encroachment (Fig. 123) on a sector that has been lowered by erosion to the water table

Figure 125 Fenced-off area on the beach at Apollo Bay, Victoria, Australia, during the nesting season of the Hooded Plover

rendered unsuitable for egg laying and incubation by erosion, by induration, or by excessive deposition.

Fishing and shellfish harvesting from beaches is traditional in many countries, and in some places it has become necessary to impose restrictions designed to protect nearshore ecology and marine organisms from over-exploitation. In Australia there have been problems where immigrants from south-east Asia come to beaches to take excessive numbers of shellfish and deplete nearshore ecosystems which local people wish to see left undisturbed as marine reserves. Sewage pollution may enrich a coastal ecosystem, but when nutrient concentrations become excessive, and algal blooms develop, ingestion by fish and shellfish can render them poisonous. Oil pollution can have an adverse effect on the taste of fish.

Watching birds and animals such as seals, porpoises and whales is an attraction on some beaches, and locally facilities such as bird hides and viewing platforms are provided (Fig. 126). A good example of a display that has drawn large numbers of tourists to a beach is the parade of penguins seen at several localities in south-eastern Australia.

Case study: Phillip Island Penguin Parade, Australia

At various points on the coasts of Bass Strait, Australia, fairy penguins come ashore each evening throughout the year, waddling across the beach and up

Figure 126 Bird-watching hut at Blakeney Point on the east coast of England

through grassy backshore hummocks to deliver food to their young in the dune burrows (Fig. 127). The most famous of these Penguin Parades occurs in Summerland Bay, on Phillip Island, Victoria, where tourists come at dusk to watch large numbers of these birds come in from the sea, where they have been feeding during the day. In the early 1930s it was realised that this penguin population was diminishing, largely because of predation by foxes and cats, and that some kind of protection was necessary if the penguins were not to disappear completely. A Penguin Reserve was created and developed commercially as a tourist attraction. As the number of visitors increased, management became necessary to protect the penguins and their habitat at the same time as permitting visitors to see the evening parade. A concrete viewing stand was built at the back of the beach, where several hundred spectators can gather, and a Visitor Centre provides information as well as organising access. The beach is floodlit to enable people to see the penguins, but use of flashlight photography is prohibited, and rangers ensure that people do not walk on the beach while the parade is in progress.

The Penguin Parade has become an international tourist attraction. Pictures of penguins making their way up Summerland Beach can be seen in the windows of travel agencies in Tokyo, London and San Fransisco, and the income from tourism has soared. However, careful management will be necessary to perpetuate this attraction, including predator control and restrictions on public access to the beach and dunes. The penguins are

Figure 127 Penguin parade on Summerland Beach, Phillip Island, Australia. Photograph by Scancolor, Australia

vulnerable to pollution, especially oil spills from the large tankers that come and go through Bass Strait and to the oil and gas terminal in nearby Western Port Bay: a major oil spill could wipe out the Penguin Parade. Where the beach is backed by cliffed dunes, cut by storm waves, penguins can no longer find their way up to their burrows in the dunes, as they are unable to scale the steep or vertical sand cliffs. In such circumstances ramps have to be provided to maintain penguin routes (Cuttriss and Bird 1995).

SOCIAL PROBLEMS

Various conflicts have developed between beach users, especially on intensively used beaches. Conflicts between swimmers, surfers and boat users can best be resolved by delimiting zones of the beach and nearshore sea for each of these purposes, and indicating them by notices, flags, markers and buoys. If vehicles are allowed on to a beach they should be confined to delineated areas.

Many complaints are made by people using a beach. Some object to hawkers, whose activities should be licensed and controlled; others to the noise of radios, vehicles or boats, especially motor racing boats and jet skis. Litter, seaweed and pollution are resented, and there is an expectation that nearshore water quality will be good enough for swimming. There are objections to dense growth of seagrasses or seaweeds in shallow nearshore areas, which impede swimming and boating, make the water less attractive, and also generate heaps of washed-up and decaying smelly vegetation on the beach, causing a sharp increase in the populations of flies and other annoying insects.

Camping on beaches is tolerated in many countries, especially away from seaside resorts and other urbanised areas, but there are often problems with unauthorised structures such as fishermen's huts and beachcombers' shacks built on beaches. Although sometimes picturesque, these can result in litter, pollution and health problems.

Dogs are often a source of conflict on beaches. Some feel that they are noisy, polluting and sometimes dangerous, and should be excluded from the beach, at least during the season of swimming and intensive beach recreation, while others accept that if they are kept under strict control by their owners (who may be legally required to remove dog excrement) there is less of a problem.

There are also complaints about unpleasant behaviour, including harassment, mainly of women and usually by men, on less frequented beaches, but the advent of mobile telephones, which can be used to report such behaviour promptly to the police, should diminish this. Some authorities prohibit the drinking of alcohol on beaches, partly because of unpleasant behaviour and the increased litter of cans and broken bottles.

Beaches generally offer no shade from the sun or shelter from wind and rain, although it is often possible to retreat to vegetated areas or shelter in structures on the backshore. Groynes may provide shelter from alongshore winds. Sandy beaches are much preferred for recreation, and there are sometimes demands that shells and stones be removed from otherwise sandy beaches. Where the beaches are of shingle many people find the smoothness and solidity of concrete or stone sea walls and groynes preferable to the gravel of the beach as places to sit or sunbathe.

The sociology of beaches deserves more attention than it has received. Deliberate segregation of racial groups was evident in South Africa during the apartheid era, but observations of beach behaviour suggest that there is a tendency for beach users to segregate themselves on the basis of race and class. Private beaches which charge entry fees in the United States afford one means of socio-economic selection, but accessibility is a key factor: in general, few people have the time and resources to get to the more remote beaches, especially on islands (West and Heatwole 1979). Aware of this, New York State endeavoured to improve public transport so that more people, especially from disadvantaged social classes, could use the beaches (Heatwole and West 1980).

BEACH HAZARDS

Incidental reference has been made to various hazards on the beach and in nearshore waters. Beaches are exposed environments, especially in storms, when strong winds and large waves can endanger people walking along the shore. During electrical storms there is a risk of being struck by lightning on an open beach. There is also a risk of being injured or swept into the sea by sudden large waves, including tsunamis generated by earthquakes or submarine eruptions. Clothes and equipment can be washed away by waves or the rising tide. Other hazards on the beach include sunburn, which is now recognised as a serious problem because of the high incidence of skin cancer in countries where beach recreation is popular. Studies have also emphasised the physiological damage to skin and eyes resulting from ultraviolet reflection from light-coloured beach sand (Kawanishi et al. 1993) and also glitter reflection from the sea (Kawanishi et al. 1994). Beach umbrellas and windshelters brought on to the beach to give shelter from sun, wind, blowing sand and rain can become dangerous when they are dislodged and blown around by strong, gusty winds, and should be firmly tethered. Beach users can also be injured by dangerous litter such as broken glass or syringes.

Rock falls and landslides on cliffs and steep coasts backing beaches are a hazard, and each year people are killed or injured in cliff accidents. The existence of a cliff indicates that erosion has occurred, but it is often difficult to assess instability and the degree of risk. Cliffs cut into soft formations frequently collapse, but the risk may be less obvious where the coastal rock outcrops appear coherent and stable, yet have occasional sudden falls. Scarborough Bluffs, on the shores of Lake Erie, in Canada, are clay cliffs subject to cyclic instability (Quigley and De Nardo 1980), and a recent fatal accident here has prompted the Canadian government to make a national review of cliff hazards.

Cliff recession is often preceded by undercutting caused largely by wave scour, so that a basal notch is a sign that collapse may be imminent. Disintegration of a cliffed rock formation is aided by the presence of joints, faults and bedding planes, and where the dip is seaward, especially where permeable strata rest upon an impermeable basement and are loosened and undermined by seaward seepage down the interface, slumping is frequent. Seepage zones in the cliff face may indicate this possibility. Cracks and fissures are an obvious hint of instability, particularly where roots of trees and shrubs penetrate and widen them, eventually dislodging rock masses. Hammering and digging cliff faces by people hunting for fossils and mineral specimens increases the risk of cliff falls, and there have been accidents during geological field excursions because of this, particularly where there are numerous sites of geological importance, as on the Dorset coast in England. Steep sand cliffs on dunes undercut by the sea may also collapse suddenly on to the beach, burying people who were scrambling up them or sitting below them. It is sometimes necessary to construct diversionary fencing to keep people away from cliff-base beach areas that are particularly dangerous, and many coastal authorities have put up notices and fences to remind people of something that should be obvious: that cliffs are hazardous (Fig. 128) (Bird 1994).

Figure 128 The hazard of undercut, slumping cliffs at Sandringham, Port Phillip Bay, Australia, has led to the erection of warning signs and the building of a fence to keep beach users away from the base of the cliff. The fence was damaged by storm waves

A surprising number of people are incapacitated by diving off piers, groynes or rocky outcrops into shallow water, and incurring severe neck and spinal injuries. Some thousands of people drown in the sea each year, mostly in nearshore areas off beaches, and it is remarkable how many people who cannot swim venture in to or on to the sea. On Australian surf beaches rip currents are a major hazard, and 89% of bathers rescued in New South Wales had been caught in them (Short and Hogan 1994). There is a risk that people afloat on rafts, lilos, rubber boats and other beach toys will be tipped into the sea by sudden large waves or swept offshore by rip currents, ebbing tides or strong winds blowing seaward. Swimmers may also encounter swift outflows from rivers and drains which carry them rapidly out to sea, and other hazards such as deep holes, unexpected reefs, broken glass and jagged rocks. Occasionally on oceanic surf beaches a sudden change in wind and wave direction results in swimmers being swept out to sea by a strong current. This is often attributed to 'sand bar collapse', but usually the explanation is that with the change in wave incidence the nearshore water circulation begins to carry water over a low sector of the sand bar, cutting out a channel through which new rip currents suddenly flow seaward (p. 28). In each case it may be necessary to erect notices warning of local hazards.

Some beaches and nearshore shallows become strewn with trees and

Figure 129 Driftwood strewn on the beach at Somers, Westernport Bay, Australia, after a storm in November 1994

branches during a storm, when scrub and woodland has been undermined by cliffing of backshore dunes (Fig. 129). Driftwood can also be produced where logging of hinterland forests yields timber washed down to the coast by rivers and floated alongshore, as on parts of the Pacific coast of North America (Fig. 130) and in New Zealand. Such driftwood can become a hazard, particularly to people swimming or diving in the nearshore shallows, and it may be necessary to remove it from heavily used recreational beaches. On some coasts it is piled up and burned, but this can lead to fire damage to backshore vegetation. It may be more usefully deposited on the backshore as a means of increasing protection from storm wave erosion, or incorporated in developing foredunes.

Encounters with harmful organisms such as mosquitoes, sandflies and other insects, as well as aggressive birds, such as terns defending their nesting sites, are a nuisance on beaches, and occasionally injurious. Other harmful marine animals encountered in nearshore areas off beaches include sharks, stinging jellyfish, stingrays, crocodiles and sea snakes. Many tropical seaside resorts find it necessary to provide fenced or netted swimming enclosures, at least in the season when harmful organisms are abundant. In Australia, Sydney Harbour has a hungry shark population, which is a threat to swimmers using harbourside beaches, while in Queensland and the Northern Territory signs warn of the dangers of stinging jellyfish and bottles of white vinegar are provided as an antidote.

Figure 130 Timber strewn on Birch Bay, Washington State, USA

Beaches on coasts that have been used for military training, or coasts of countries that are, or have recently been, at war may be dangerous because of the presence of unexploded bombs and shells or land mines. Containers with poisonous chemicals or radioactive substances have been washed up on beaches, and wrecked ships can become a beach hazard.

The various problems arising from beach use may require that beach management agencies put up notices at beach access points indicating local regulations (Fig. 131) and defining zones in which specific activities are either permitted or prohibited (Fig. 132). Recognition of beach hazards has led to the introduction of various beach safety regulations and the provision of security and life-saving patrols at many seaside resorts, particularly in the United States, Australia and New Zealand. Surf Life Saving Australia is a national organisation co-ordinating the efforts of 255 Life Saving Clubs and professional lifeguards who patrol 300 beaches and make over 10 000 rescues each year. This organisation was one of the sponsors of the Beach Safety and Management Program, which is documenting coastal hazards and their impacts on public safety on more than 7000 Australian ocean beaches. It has developed a database for every beach, showing location, access, nature, physical characteristics, facilities, use, and beach and surf conditions, together with an assessment of risk levels (on a scale 1 to 10) and a determination of resources required to maintain adequate levels of public safety on each beach (Short et al. 1993).

Figure 131 Sign indicating regulations for beach use, Port Phillip Bay, Australia

Vehicles on beaches

Cars and motor cycles can be driven (and parked) on beaches where the sand is firm enough. Vehicle access to beaches, where permitted, should use solid ramps, because wheeled traffic (including cars, four-wheel drive vehicles, beach buggies and motor cycles) leads to erosion along tracks made through vegetated dunes or down steep coastal slopes. Once on the beach, vehicles driven along the shore have little impact in the area washed by waves at high tide, but can be a hazard to other beach users, and restrictions are necessary to avoid accidents.

Some beaches have been used for motor racing, as at Pendine Sands in Wales, and reference has been made to the sport of sand yachting (p. 220).

Recreational vehicles can be driven along some beaches, particularly on coasts where road access is limited, as on Fraser Island in south-east

Figure 132 Notice warning of the nudist hazard on the beach at Studland, southern England

Queensland or the shores of Brunei. The beach may be firm enough to be used by buses and light aircraft, as on Four Mile Beach in North Queensland, and commercial aircraft land on the sandy beach at low tide on the Hebridean island of Barra.

Restrictions have been placed on vehicles using beaches in many countries, not only because of damage to dunes, cliffs and backshore vegetation at points where such vehicles are driven on or off a beach, but also because of the risk of accidents. In Queensland, the Beach Protection Agency requires that such vehicles use designated access points, that they drive only on the wet sand zone between the swash limits at high and low tides and keep well clear of other beach users.

Figure 133 Cars on the beach at low tide, Rømø, Denmark

On North Sea coast beaches in Germany and Denmark there are problems where cars are taken on to the beach at low tide (Fig. 133) and parked. Sometimes the owners return from a walk to find them awash. It is quite easy to 'bog' vehicles in soft sand, and if they are not quickly retrieved they may be submerged by a rising tide. Many beach users object to vehicles on beaches, and some coastal management agencies regard car and motor cycle traffic as incompatible with beach environments, and prohibit it entirely, but access for beach patrol cars and emergency vehicles such as ambulances may occasionally be necessary.

SURVEYS OF BEACH USE

There have been various surveys of beach use, usually by interview and questionnaire, but most of them remain as unpublished reports, which are difficult to obtain even locally. Some surveys consider beaches as part of a wider environment, such as a National Park (Cofer-Shabica et al. 1990).

Beach surveys face a number of difficulties. The timing of such surveys is important, for there are variations in the number and kind of people using a beach, and in the activities pursued, with weather and sea conditions, day of the week, time of day, and peak and off seasons. There can be reluctance on the part of beach visitors to take part: one survey of recreation of surf beaches in Australia had a surprisingly high proportion (78.3%) of female respondents, presumably because most of the males were out in the sea riding their surfboards at the time (Tower and Kain 1993).

Some surveys focus on the sociology of beach users, including their age, gender and group characteristics, education, occupation and income. Others are concerned with the timing and frequency of their visits, where they have come from, by what mode of transport, why they chose a particular beach, what activities they pursued on the beach, how much they spend locally, and what improvements they would like to see.

Evaluation of the recreational use of beaches often examines the costs and benefits of beach use, generally as a basis for seeking larger expenditure nationally or locally on beach management. There are often questions exploring how the costs of maintaining beach use facilities (including cleaning, life-saving and policing) should be met. Coastal municipalities and beach management organisations may argue that the beach is also used by people coming from outside their region, and that a contribution should come from other (e.g. national) sources, or that access fees should be charged, or sponsorship in exchange for advertising. Others explore people's awareness of the various hazards and conflicts that have been mentioned, and their reaction and response to warning signs that attempt to reduce accidents on the beach.

CONCLUSION

Satisfactory beach management requires an understanding of the nature and dynamics of beach systems, the various physical, chemical, biological and social interactions that take place on and around them, and the aims and perceptions of people who come to use them. Beach managers have the responsibility of maintaining a wholesome beach and nearshore environment, and endeavouring to provide beach users with opportunities to enjoy the various experiences they seek when they arrive, usually with high expectations, on the shore. It is impossible to please all beach users all of the time, but many conflicts can be resolved by judicious zoning of beach and nearshore activities.

Beaches have been the focus of strong interest, attention and activity over the past two centuries, since the days when Dr. Richard Russell first advocated a visit to the seaside town of Brighthelmstone (now Brighton, on the south coast of England) on medical grounds. The rise of seaside resorts and beach holidays followed, and shows little sign of abating. Until it does, beaches will continue to be seen as playgrounds at the edge of land and sea, places of opportunity, entertainment, exercise and enjoyment, and beach management will be necessary to optimise these values.

References

Adams, J.W.R. (1979) Rebuilding the beaches of Florida. *Shore and Beach*, **47**: 3–6.

Anders, F.J. and Hansen, M. (1990) *Beach and borrow site investigation for a beach nourishment at Ocean City, Maryland.* US Army Corps of Engineers, Waterways Experiment Station, Coastal Engineering Research Center, Technical Report 90–5.

Ashley, G.M., Halsey, S.D. and Farrell, S.C. (1987) A study of beach fill longevity: Long Beach Island, New Jersey. *Proceedings Coastal Sediments '87*, pp. 1188–1202.

Baca, B.J. and Lankford, T.E. (1988) *Myrtle Beach nourishment project: biological monitoring*. Report to City of Myrtle Beach.

Baker, G. (1956) Sand drift at Portland, Victoria. *Proceedings of the Royal Society of Victoria*, **68**: 151–198.

Barrett, M.G. (1989) What is coastal management? In: M.G. Barrett (ed.) *Coastal Management. Proceedings of the Conference organised by the Institution of Civil Engineering in Bournemouth*, 9–11 May 1989. Telford, London, pp. 1–9.

Barrett, M.G. (1993) *Coastal Zone Planning and Management.* Thomas Telford, London.

Barth, M.G. and Titus, J.G. (1984) *Greenhouse Effect and Sea Level Rise.* Van Nostrand Reinhold, New York.

Bascom, W.N. (1951) Relationship between sand size and beach face slope. *Transactions of the American Geophysical Union*, **32**: 866–874.

Basinski, T. (1994) Protection of Hel Peninsula. In: K. Rotnicki (ed.) *Changes of the Polish Coastal Zone*. Quaternary Research Institute, Adam Mickiewicz University, Poznan, pp. 53–56.

Berg, D.W. and Duane, D.B. (1968) Effect of particle size and distribution on stability of artificially filled beach, Presque Isle Peninsula, Pennsylvania. *Proceedings of the 11th Conference on Great Lakes Research*, pp. 161–178.

Beven, S.M. (ed.) (1985) *Shingle Beaches Renourishment and Recycling*. Hydraulics Research, Wallingford, UK.

Bird, E.C.F. (1971) The origin of beach sediments on the North Queensland coast. *Earth Science Journal*, **5**: 95–105.

Bird, E.C.F. (1978) The nature and source of beach materials on the Australian coastline. In: J.L. Davies and M.A.J. Williams (eds.) *Landscape Evolution in Australasia*, Australian National University, Canberra, pp. 144–157.

Bird, E.C.F. (1981) Beach erosion problems at Wewak, Papua New Guinea. *Singapore Journal of Tropical Geography*, **2**: 9–14.

Bird, E.C.F. (1984) *Coasts*, 3rd edn. Australian National University Press and Blackwell, Oxford, 320 pp.

Bird, E.C.F. (1985a) *Coastline Changes: A Global Review*. Wiley, Chichester.

Bird, E.C.F. (1985b) Recent changes on the Somers-Sandy Point coastline. *Proceedings of the Royal Society of Victoria*, **97**(3): 115–128.
Bird, E.C.F. (1987) The effects of quarry waste disposal on beaches on the Lizard Peninsula, Cornwall. *Journal of the Trevithick Society*, **14**: 83–92.
Bird, E.C.F. (1988) The future of the beaches. In: R.L. Heathcote (ed.) *The Australian Experience*. Longman Cheshire, Melbourne, pp. 163–177.
Bird, E.C.F. (1990a) Artificial beach nourishment on the shores of Port Phillip Bay, Australia. *Journal of Coastal Research*, Special Issue, **6**: 55–68.
Bird, E.C.F. (1990b) Reed encroachment on Estonian beaches. *Proceedings of the Estonian Academy of Sciences*, **39**: 7–11.
Bird, E.C.F. (1991) Changes on artificial beaches in Port Phillip Bay, Australia. *Shore and Beach*, **59**: 19–27.
Bird, E.C.F. (1993a) *The Coast of Victoria*. Melbourne University Press, Melbourne.
Bird, E.C.F. (1993b) *Submerging Coasts: The Effects of a Rising Sea Level on Coastal Environments*. Wiley, Chichester.
Bird, E.C.F. (1993c) Geomorphological aspects of surfing in Victoria, Australia. In: P.P. Wong (ed.) *Tourism vs Environment: The Case for Coastal Areas*. Kluwer, The Netherlands, pp. 11–18.
Bird, E.C.F. (1994) Cliff hazards and coastal management. *Journal of Coastal Research*, Special Issue, **12**: 299–309.
Bird, E.C.F. (1996) Lateral grading of beach sediments: a commentary. *Journal of Coastal Research*, in press.
Bird, E.C.F. and Christiansen, C. (1982) Coastal progradation as a by-product of human activity: an example from Hoed, Denmark. *Geografisk Tidsskrift*, **82**: 1–4.
Bird, E.C.F. and Fabbri. P. (1993) Geomorphological and historical changes on the Argentina Delta, Ligurian coast, Italy. *GeoJournal*, **29**: 428–429.
Bird, E.C.F. and Jones, D.J.B. (1988) The origin of foredunes on the coast of Victoria, Australia. *Journal of Coastal Research*, **4**: 181–192.
Bird, E.C.F. and May, V.J. (1976) *Shoreline changes in the British Isles during the past century*. IGU Working Group Paper, Division of Geography, Bournemouth College of Technology.
Bird, E.C.F. and Ongkosongo, O.S.R. (1980) *Environmental Changes on the Coasts of Indonesia*. United Nations University, Tokyo.
Bird, E.C.F. and Rosengren, N.J. (1984) The changing coastline of the Krakatau Islands, Indonesia. *Zeitschrift Geomorphologie*, **28**: 346–366.
Bird, E.C.F., Dubois, J.P. and Iltis , J.A. (1984) *The Impacts of Opencast Mining on the Rivers and Coasts of New Caledonia*. United Nations University, Tokyo.
Black, D.W., Donnelley, L.P. and Settle, R.F. (1988) *An Economic Analysis of Beach Renourishment for the State of Delaware*. University of Delaware.
Bodéré, J.C. (1979) Le rôle essential des débacles glacio-volcaniques dans l'évolution recente des côtes sableuses en voie de progradation du sud-est d'Islande. In: A. Guilcher (ed.) *Les Côtes Atlantiques de l'Europe*. University of Western Brittany, Brest, pp. 55–64.
Bodge, K.R. (1991) Damage benefits and cost sharing for shore protection projects. *Shore and Beach*, **59**: 11–18.
Bodge, K.R. (1992) Representing equilibrium beach profiles with an exponential expression. *Journal of Coastal Research*, **8**: 47–55.
Borowca, M. and Rotnicki, K. (1994) Intensity, directions and balance of aeolian transport on the beach barrier and the problem of sand nourishment. In: K. Rotnicki (ed.) *Changes of the Polish Coastal Zone*. Quaternary Research Institute, Adam Mickiewicz University, Poznan, pp. 108–115.

Bourman, R.P. (1990) Artificial beach progradation by quarry waste disposal at Rapid Bay, South Australia. *Journal of Coastal Research*, Special Issue, **6**: 69–76.
Brampton, A.H. (1977) *A Computer Model for Wave Refraction*. Report 172, Hydraulics Research Station, Wallingford, UK.
Brampton, A.H. and Motyka, J.M. (1983) The effectiveness of groynes, *Shoreline Protection Conference*. Telford, London, pp. 151–156.
Brampton, A.H. and Motyka, J.M. (1987) Recent examples of mathematical models of UK beaches. *Proceedings Coastal Sediments '87*, pp. 515–530.
Bremner, J.M. (1985) Southwest Africa/Namibia. In: E.C.F. Bird and M.L. Schwartz (eds.) *The World's Coastline*. Van Nostrand Reinhold, New York, pp. 645–651.
Brown, A.C. and McLachlan, A. (1990) *Ecology of Sandy Shores*. Elsevier, Amsterdam.
Brown, M.J.F. (1974) A development consequence: disposal of mining waste on Bougainville, Papua New Guinea. *Geoforum*, **18**: 19–27.
Brownlie, W.R. and Brown, W.M (1978) Effects of dams on beach sand supply. *Proceedings Coastal Zone '78*, pp. 2273–2278.
Bruun, P. (1954) *Coast Erosion and the Development of Beach Profiles*. Beach Erosion Board, Technical Memoir 44.
Bruun, P. (1962) Sea level rise as a cause of shore erosion. *Journal of Waterways and Harbors Division, Proceedings of the American Society of Civil Engineers*, **88**: 117–130.
Bruun, P. (1989) The coastal drain: what can it do or not do? *Journal of Coastal Research*, **5**(1): 123–125.
Bruun, P. (1990) Beach nourishment – improved economy through better profiling and backpassing from offshore sources. *Journal of Coastal Research*, **6**: 265–277.
Bryant, E.A. (1985) Rainfall and beach erosion relationships, Stanwell Park, Australia, 1895–1980: worldwide implications for coastal erosion. *Zeitschrift für Geomorphologie*, Supplementband **57**: 51–65.
Campbell, J.F. and Moberley, R. (1984) USA–Hawaii. In: H.J. Walker (ed.) *Artificial Structures and Shorelines*. Louisiana State University, pp. 1–4.
Campbell, T.J. and Beachler, K. (1984) The 1983 beach restoration of Pompano and Lauderdale by the sea. *Shore and Beach*, **52**: 21–26.
Campbell, T.J. and Spadoni, R.H. (1982) Beach restoration - an effective way to combat erosion on the southeast coast of Florida. *Shore and Beach*, **20**: 11–12.
Carr, A.P. and Blackley, M.W.L. (1974) Ideas on the origin and development of Chesil Beach, Dorset. *Proceedings of the Dorset Natural History and Archaeology Society*, **95**: 1–9.
Carter, R.W.G. (1988) *Coastal Environments*. Academic Press, London.
CERC (Coastal Engineering Research Center) (1987) *Shore Protection Manual*. 4th edn. US Army, Vicksburg.
Chapman, D.M. (1978) Beach behaviour following nourishment, Lower Gold Coast, Queensland. *Search*, **9**: 460–462.
Chappell, J., Eliot, I.G., Bradshaw, E. and Lonsdale, E. (1979) Experimental control of beach face dynamics by water table pumping. *Engineering Geology*, **14**: 29–41.
Chen Jiyu, Liu Cangzi and Yu Zhiying (1985) China. In: E.C.F. Bird and M.L. Schwartz (eds.) *The World's Coastline*. Van Nostrand Reinhold, New York, pp. 813–822.
Chill, J., Butcher, C. and Dyson, W. (1989) Beach nourishment with fine sand at Carlsbad, California. *Proceedings Coastal Zone '89*, pp. 2091–2103.
Chou, I.B., Powell, G.M. and Winton, T.C. (1983) Assessment of beach fill performance by excursion analysis. *Proceedings Coastal Zone '83*, pp. 2361–2386.

Christiansen, C., Christoffersen, H., Dalsgaard, J. and Nornberg, P. (1981) Coastal and nearshore changes correlated with die-back in eel grass (*Zostera marina*). *Sedimentary Geology*, **28**: 163–173.
Christiansen, H. (1977) Economic profiling of beach fills. *Proceedings Coastal Sediments '77*, pp. 1042–1048.
Ciprani, L.E., Dreoni, A.M. and Pranzini, E. (1992) Nearshore morphological and sedimentological evolution induced by beach restoration: a case study. *Bollettino di Oceanologica Teorica ed Applicata*, **10**: 279–295.
Clark, A.R. (1988) The use of Portland Stone armour in coastal protection and sea defence works. *Quarterly Journal of Engineering Geology*, **21**: 113–136.
Clayton, K.M. (1988) Sediment input from the Norfolk cliffs, eastern England – a century of coast protection and its effect. *Journal of Coastal Research*, **5**: 433–442.
Clayton, K.M. (1989) Implications of climatic change. In: Coastal Management. *Proceedings of the Conference organised by the Institution of Civil Engineering in Bournemouth*, 9–11 May 1989. Telford, London, pp. 165–176.
Clayton, T.D. (1989) Artificial beach nourishment on the Pacific shore: a brief overview. *Proceedings Coastal Zone '89*, pp. 2033–2045.
Cofer-Shabica, S., Snow, R.E. and Noe, F.P. (1990) Formulating policies using visitor perception of Biscayne National Park. In: P. Fabbri (ed.) *Recreational Uses of Coastal Areas*. Kluwer, Dordrecht, pp. 235–256.
Coughlan, P.M. (1989) Noosa Beach – Coastal engineering works to mitigate the erosion problem. *9th Annual Conference Coastal & Ocean Engineering*, pp. 198–203.
Craig, A.K. and Psuty, N.P. (1968) The Paracas Papers. *Studies in Marine Desert Ecology*, **1**: 44–77.
Craig-Smith, S.J. (1973) *Sand Injection in the United Kingdom: The Position in 1973*. East Anglian Coastal Study Report, No. 5. University of East Anglia.
Cressard, A.P. and Augris, C. (1982) Etude des phénomènes d'érosion cotière liès a l'extraction de materiaux sur le plateau continental. *Proceedings 4th Conference International Association of Engineering Geology*, vol. 7, pp. 203–211.
Cruz, O., Coutinho, P.N., Duarte, G.M., Gomes, A. and Muehe, D. (1985) Brazil. In: E.C.F. Bird and M.L. Schwartz (eds.) *The World's Coastline*. Van Nostrand Reinhold, New York, pp. 85–91.
Cullen, P. and Bird, E.C.F. (1980) *The Management of Coastal Sand Dunes in South Australia*. Geostudies, Melbourne.
Cunningham, R.T. (1966) Evaluation of Bahaman oolitic aragonite sand for Florida beach nourishment. *Shore and Beach*, **34**: 18–21.
Cuttriss, L. and Bird, E.C.F. (1995) *500 Million Years on Phillip Island*. Bass Coast Shire Council, Victoria, Australia.
Davies, D.K., Etheridge, P.G. and Berg, R.R. (1971) Recognition of barrier environments. *Bulletin American Association of Petroleum Geologists*, **55**: 550–565.
Davies, J.L. (1974) The coastal sediment compartment. *Australian Geographical Studies*, **12**: 139–151.
Davies, J.L. (1980) *Geographical Variation in Coastal Development*. Longman, London.
Davis, G.A., Hanslow, D.J., Herbert, K. and Nielsen, P. (1992) Gravity drainage: a new method of beach stabilisation through drainage of the watertable. *Proceedings 23rd International Conference on Coastal Engineering*, Venice.
Davis, R.A. (1983) *Depositional Systems*. Prentice Hall, New York.
Davison, A.T. (1992) The National Flood Insurance Mitigation and Erosion

Management Act of 1991: Background and Overview. *Proceedings National Conference on Beach Preservation Technology '92*, pp. 77–85.
Davison, A.T., Nicholls, R.J. and Leatherman, S.P. (1992) Beach nourishment as a coastal management tool: an annotated bibliography on developments associated with the artificial nourishment of beaches. *Journal of Coastal Research*, **8**: 984–1022.
Dean, R.G. (1973) Heuristic models of sand transport in the surf zone. *Proceedings of a Conference on Engineering Dynamics in the Surf Zone*, pp. 208–214.
Dean, R.G. (1983) Principles of beach nourishment. In: P.D. Komar (ed). *Handbook of Coastal Processes and Erosion*. C.R.C. Press, Boca Raton, FL, pp. 53–84.
Dean, R.G. (1987a) Additional sediment inputs to the nearshore region. *Shore and Beach*, **55**: 76–81.
Dean, R.G. (1987b) Realistic economic benefits from beach nourishment. *Proceedings Coastal Sediments '87*, pp. 1558–1572.
Dean, R.G. (1991) Equilibrium beach profiles: characteristics and applications. *Journal of Coastal Research*, **7**: 53–84.
De Lange, W.P. and Healy, T.R. (1990) Renourishment of a flood-tidal delta adjacent beach, Tauranga Harbour, New Zealand. *Journal of Coastal Research*, **6**: 627–640.
Delft Hydraulics Laboratory (1970) *Gold Coast, Queensland, Australia – Coastal Erosion and Related Problems*. Delft.
Delft Hydraulics Laboratory (1987) *Manual on Artificial Beach Nourishment*. Centre for Civil Engineering Research, Codes and Specifications, Rijkwaterstaat.
Delft Hydraulics Laboratory (1992) *Southern Gold Coast Littoral Sand Supply*. Delft.
Dette, H.H. (1977) Effectiveness of beach deposit nourishment. *Proceedings Coastal Sediments '77*, pp. 211–227.
Dette, H.H. (1990) Offshore sand dredge and delivery systems. *Proceedings 3rd Annual National Beach Preservation Technology Conference*, pp. 378–393.
Dixon, K. and Pilkey, O.H. (1989) Beach replenishment on the US Coast of the Gulf of Mexico. *Proceedings Coastal Zone '89*, pp. 2007–2020.
Dolan, R. (1971) Coastal landforms: crescentic and rhythmic. *Bulletin Geological Society of America*, **82**: 177–180.
Doody, P. (ed.) (1985) *Sand Dunes and their Management*. Nature Conservancy Council, Peterborough, UK.
Downie, K.A. and Saaitink, H. (1983) An artificial cobble beach for erosion control. *Proceedings Coastal Structures '83*, pp. 846–859.
Drew, P. (1994) *The Coast Dwellers*. Penguin, Australia.
Drewe, R. (1987) *The Bodysurfers*. Picador, Sydney.
Dubois, R.N. (1977) Predicting beach erosion as a function of rising water level. *Journal of Geology*, **85**: 470–476.
Dutton. G. (1985) *Sea, Surf and Sand: The Myth of the Beach*. Oxford University Press, Melbourne.
Eddison, J. (1983) The evolution of the barrier beaches between Fairlight and Hythe. *Geographical Journal*, **149**: 39–53.
Eitner, V. and Ragutzki, G. (1994) Effects of artificial beach nourishment on nearshore sediment distribution (Island of Nordeney, Southern North Sea). *Journal of Coastal Research*, **10**: 637–650.
Emery, K.O. (1960) *The Sea off Southern California*. Wiley, New York.
Empsall, B. (1989) Workington ironworks reclamation. In: Coastal Management.

Proceedings of the Conference organised by the Institution of Civil Engineering in Bournemouth, 9–11 May 1989. Telford, London, pp. 279–292.
Evangelista, S., La Monica, G.B. and Landini, B. (1992) Artificial beach nourishment using fine crushed limestone gravel at Terracina (Latium, Italy). *Bollettino di Oceanologica Teorica ed Applicata*, **10**: 273–278.
Everard, C.E. (1962) Mining and shoreline evolution near St. Austell, Cornwall. *Transactions of the Royal Geological Society, Cornwall*, **19**: 199–219.
Everts, C.H. (1985) Sea-level rise effects on shoreline position. *Journal of Waterway, Port, Coastal and Ocean Engineering*, **111**: 985–999.
Everts, C.H., DeWall. A.E. and Czerniak, M.T. (1974) Behaviour of beach fill at Atlantic City, New Jersey. *Proceedings of the 14th Conference on Coastal Engineering*, pp. 1370–1388.
Fabbri, P. (1985) Coastline variations on the Po delta since 2500 B.P. *Zeitschrift für Geomorphologie*, Supplementband **57**: 155–167.
Fabbri, P. (1990) *Recreational Uses of Coastal Areas*. Kluwer, Dordrecht.
Fanos, A.M., Khafagy, A.A. and Dean, R.G. (1995) Protective works on the Nile delta coast. *Journal of Coastal Research*, **11**: 516–528.
Finkl, C.W. (1981) Beach nourishment, a practical method of erosion control. *Geo-Marine Letters*, **1**: 155–161.
Fisher, J.J. (1980) Shoreline erosion, Rhode Island and North Carolina coasts. In: M.L. Schwartz and J.J. Fisher (eds.) *Proceedings of the Per Bruun Symposium*, University of Rhode Island, pp. 32–54.
Fotheringham, D.G. and Goodwins, D.R. (1990) Monitoring the Adelaide beach system. *Proceedings 1990 Workshop on Coastal Zone Management*, Yeppoon, Queensland, pp. 118–132.
Foxley, J.C. and Shave, K. (1983) Beach monitoring and shingle recharge. *Shoreline Protection Conference*. Telford, London, pp. 163–170.
Fulford, E.T. and Grosskopf, W.G. (1989) Storm protection project design, Ocean City, Maryland. *Beach Preservation Technology '89*, pp. 239–248.
Gale, R.W. (1979) Coast and beach protection in Queensland, Australia. *Shore and Beach*, **49**: 3–6.
Galster, R.W. and Schwartz, M.L. (1990) Ediz Hook – a case study of coastal erosion and rehabilitation. *Journal of Coastal Research*, Special Issue, **6**: 103–114.
Gell, R.A. (1978) Shelly beaches on the Victorian coast. *Proceedings of the Royal Society of Victoria*, **90**: 257–269.
Gentleman, D. (1988) *Coastline*. Weidenfeld and Nicolson, London.
Goemans, T. (1986) The sea also rises: the ongoing dialogue of the Dutch with the sea. In: J.G. Titus (ed.) *Effects of Changes in Stratospheric Ozone and Global Climate*, vol. 4, US Environment Protection Agency, pp. 47–56.
Goldsmith, V. (1978) Coastal dunes. In: R.A. Davis (ed.) *Coastal Sedimentary Environments*, Springer, New York, pp. 171–236.
Golik, A. and Gertner, Y. (1992) Litter on the Israeli coastline. *Marine Environmental Research*, **33**: 1–15.
Gowlland-Lewis, M., Fulton, I. and Audas, D. (1996) *Darwin Coastal Erosion Survey*. Technical Report No. 56, Resources Conservation Branch, Department of Lands, Planning and Environment, Northern Territory, Darwin, Australia.
Grant, U.S. (1948) Influence of water table on beach aggradation and degradation. *Journal of Marine Research*, **7**(3): 655–660.
Gresswell, R.K. (1953) *Sandy Shores in South Lancashire*. University of Liverpool Press, Liverpool.
Griggs, G.B. (1990) Littoral drift impoundment and beach nourishment in northern Monterey Bay, California. *Journal of Coastal Research*, Special Issue, **6**: 115–126.

Grove, A.T. (1953) The sea flood on the coasts of Norfolk and Suffolk. *Geography*, **38**: 164–170.

Grove, R.S., Sonu, C.J. and Dykstra, D.H. (1987) Fate of massive sediment injection on a smooth shoreline at San Onofre, California. *Proceedings Coastal Sediments '87*, pp. 531–538.

Guerin, B. (1984) Beach restoration, Mentone, Port Phillip Bay, Victoria. In: M.J. Knight, E.J. Minty and R.B. Smith (eds.) *Case Studies in Engineering Geology, Hydrogeology and Environmental Geology*. Special Publication, Geological Society of Australia, vol. 11, pp. 398–410.

Guilcher, A. (1985) Angola. In: E.C.F. Bird and M.L. Schwartz (eds.) *The World's Coastline*. Van Nostrand Reinhold, New York, pp. 639–643.

Guza, R.T. and Inman, D.L. (1975) Edge waves and beach cusps. *Journal of Geophysical Research*, **80**: 2997–3012.

Hales, L.Z., Byrnes, M.R. and Dowd, M.W. (1991) Numerical modelling of storm-induced beach erosion, Folly Beach, South Carolina. *Proceedings Coastal Zone Conference '91*, pp. 495–509.

Hall, J.V. (1952) Artificially constructed and nourished beaches in coastal engineering. *Proceedings 3rd Coastal Engineering Conference*, pp. 119–133.

Hallégouet, B. and Guilcher, A. (1990) Moulin Blanc artificial beach, Brest, Western Brittany, France. *Journal of Coastal Research*, Special Issue, **6**: 17–20.

Hansen, M.E. and Byrnes, M.R. (1991) Development of optimum beach fill design cross section. *Proceedings Coastal Sediments '91*, pp. 2067–2080.

Hanslow, D.J., Davis, G.A., Bowen, R., Reed, A., Sario, R., Herbert, K. and Nielsen, P. (1993) Morphological effect of artificial beach drainage. *Proceedings 17th Australian Conference on Coastal and Ocean Engineering*, pp. 711–713.

Hanson, H. (1987) *Genesis – A Generalised Shoreline Change Numerical Model for Engineering Use*. Lund University, Paper 1007, Lund, Sweden.

Hardisty, J. (1990) *Beaches: Form and Process*. Unwin Hyman, London.

Hardisty, J. (1994) Beach and nearshore sediment transport. In: K. Pye (ed.) *Sediment Transport and Depositional Processes*. Blackwell, Oxford, pp. 219–255.

Hayes, M.O. (1972) Forms of sediment accumulation in the beach zone. In: R.E. Meyer (ed.) *Waves on Beaches*. Academic Press, New York, pp. 297–356.

Healy, T.R. (1977) Progradation at the entrance to Tauranga Harbour, Bay of Plenty. *New Zealand Geographer*, **33**: 90–91.

Healy, T.R., Kirk, R.M., and De Lange, W.P. (1990) Beach renourishment in New Zealand. *Journal of Coastal Research*, Special Issue, **6**: 77–90.

Heatwole C.A. and West, N.C. (1980) Mass transit and beach use in New York City. *Geographical Review*, **70**: 210–217.

Hecock, R. (1970) Recreation behaviour patterns as related to site characteristics of beaches. *Journal of Leisure Research*, **2**: 237–250.

Herron, W.J. (1980) Artificial beaches in southern California. *Shore and Beach*, **48**: 3–12.

Hesp, P.A. (1984) Foredune formation in southeast Australia. In: B.G. Thom (ed.) *Coastal Geomorphology in Australia*. Academic Press, Sydney, pp. 69–97.

Heydorn, A.E.F. and Tinley, K.L. (1980) *Estuaries of the Cape, Part I. Synopsis of the Cape Coast*. National Research Institute, Oceanology, Stellenbosch, South Africa.

Hobson, R.D. (1977) Sediment handling and beach fill design. *Proceedings Coastal Sediments '77*, pp. 167–180.

Hough, J.L. and Menard, H.W. (1956) *Finding Ancient Shorelines*. Society of Economic Palaeontologists and Mineralogists, Special Publication 3, Tulsa, OK.

Houston, J.R. (1990) Discussion of editorial by O.H. Pilkey (1990) and paper by Leonard et al. (1990). *Journal of Coastal Research*, **6**: 1023–1035.

Houston, J.R. (1991a) Rejoinder to Discussion of Pilkey and Leonard (1990). *Journal of Coastal Research*, **7**: 565–577.

Houston, J.R. (1991b) Ocean City, Maryland, beachfill performance. *Shore and Beach*, **59**: 15–24.

Houston, J.R. (1995) Beach replenishment. *Shore and Beach*, **63**: 21–24.

Howell, S. (1974) *The Seaside*. Studio Vista, London.

Hydraulics Research Station (Wallingford, UK) (1970) *Colliery Waste Dumping on the Durham Coast*. Report No. 521.

Ibe, A.C., Awosika, L.F., Ibe, C.E. and Inegbedion, L.E. (1991) Monitoring of the 1985–86 beach nourishment project at Bar Beach, Victoria Island, Lagos, Nigeria. *Proceedings Coastal Zone '91*, pp. 534–552.

Ignatov, Y.I., Kaplin, P.A., Lukyanova, S.A. and Solovieva, G.D. (1993) Evolution of Caspian Sea coasts under conditions of sea level rise. *Journal of Coastal Research*, **9**: 104–111.

Ingle, D. (1966) *The Movement of Beach Sand*. Elsevier, Amsterdam.

Jackson, J.M. (1985) Uruguay. In: E.C.F. Bird and M.L. Schwartz (eds.) *The World's Coastline*. Van Nostrand Reinhold, New York, pp. 77–84.

Jackson, N.L. and Nordstrom, K.F. (1994) The mobility of beach fill in front of a seawall on an estuarine shoreline, Cliftwood Beach, New Jersey, USA. *Ocean & Coastal Management*, **23**: 149–166.

Jacobsen, E.E. and Schwartz, M.L. (1981) The use of geomorphic indicators to determine the direction of net shore-drift. *Shore and Beach*, **49**: 38–43.

James, W.R. (1974) Beach fill stability and borrow material structure. *Coastal Engineering, 1974 Proceedings*, pp. 1334–1344.

James, W.R. (1975) *Techniques in Evaluating Stability of Borrow Material for Beach Nourishment*. US Army Coastal Engineering Research Center, Technical Manual 60.

Jansen, W.A. (1985) Shore stabilization and beach nourishment, Illinois Beach State Park. *Shore and Beach*, **53**: 3–6.

Jelgersma, S. (1975) Netherlands. In: E.C.F. Bird and M.L. Schwartz (eds.) *The World's Coastline*. Van Nostrand Reinhold, New York, pp. 343–352.

Johnson, J.W. (1956) Dynamics of nearshore sediment movement. *Bulletin American Association of Petroleum Geologists*, **40**: 2211–2232.

Johnson, J.W. (1959) The littoral drift problem at shoreline harbours. *Journal of the Waterways and Harbors Division, Proceedings of the American Society of Civil Engineers*, **124**: 525–555.

Jolliffe, I.P. (1961) The use of tracers to study beach movements and the measurement of littoral drift by a fluorescent technique. *Revue de Géomorphologie Dynamique*, **12**: 81–95.

Jolliffe, I.P. (1964) An experiment designed to compare the relative rates of movement of beach pebbles. *Proceedings, Geologists' Association*, **75**: 67–86.

Jones, E. (1975) Jamaica. In: E.C.F. Bird and M.L. Schwartz (eds.) *The World's Coastline*. Van Nostrand Reinhold, New York, pp. 173–180.

Jones, J.C.E. and Schafter, I.F. (1986) Case studies of beach nourishment in Port Phillip Bay, Victoria, Australia. *Manual on Artificial Beach Nourishment*, Delft Hydraulics Laboratory, pp. 26–36.

Jones, M.A.J. and Bacon, P.R. (1980) Beach tar contamination in Jamaica. *Marine Pollution Bulletin*, **21**: 331–334.

Kana, T.W. (1989) Erosion and beach restoration at Seabrook Island, South Carolina. *Shore and Beach*, **57**: 3–18.

Kana, T.W. and Stevens, F.D. (1990) Beach and dune restoration following Hugo. *Shore and Beach*, **58**: 57–63.

Kaplin, P.A. (1989) Shoreline evolution during the twentieth century. In: A. Ayale-Castañares, W. Wooster and A. Yañez-Aranciba (eds.) *Oceanography 1988*. National University Press, Mexico City, pp. 59–64.

Kawanishi, T., Kadomatsu, H., Sato, K. and Watanabe, N. (1993) Basic study of ultraviolet reflection on coastal sand. *Proceedings 3rd International Offshore and Polar Engineering Conference*, pp. 211–215.

Kawanishi, T., Kumada, A. and Kato, M. (1994) Measurements of sea surface sun-glitter. *Proceedings Oceans '94 Conference*, vol. 2, pp. 706–709.

Kelletat, D. (1992) Coastal erosion and protection measures at the German North Sea coast. *Journal of Coastal Research*, **8**: 699–711.

Kerckaert, P., Paul, P.L., Roovers, A.N. and De Cand, T.P. (1986) Artificial beach nourishment on Belgian east coast. *Journal of Waterways, Port, Coastal and Ocean Engineering*, **112**: 560–571.

Kiknadze, A., Sakvarelidze, V.V., Peshkov, P. and Russo, G.E. (1990) Beach-forming process management of the Georgian Black Sea coast. *Journal of Coastal Research*, Special Issue, **6**: 33–44.

King, C.A.M. (1972) *Beaches and Coasts*. 2nd edn. Arnold, London.

Kirk, R.M. and Weaver, R.J. (1985) Coastal erosion and beach re-nourishment at Washdyke Lagoon, South Canterbury, New Zealand. *Proceedings Australasian Conference on Coastal and Ocean Engineering*, vol. 1, pp. 519–524.

Koike, K. (1985) Japan. In: E.C.F. Bird and M.L. Schwartz (eds.) *The World's Coastline*. Van Nostrand Reinhold, New York, pp. 843–855.

Koike, K. (1990) Artificial beach construction on the shores of Tokyo Bay, Japan. *Journal of Coastal Research*, Special Issue, **6**: 45–54.

Komar, P.D. (1976) *Beach Processes and Sedimentation*. Prentice-Hall, Englewood Cliffs, NJ.

Komar, P.D. (1983) *Handbook of Coastal Processes and Erosion*. CRC Press, Boca Raton, FL.

Kraus, N.C. and Pilkey, O.H. (eds.) (1988) The effects of seawalls on the beach. *Journal of Coastal Research*, Special Issue, **4**.

Krumbein, W.C. (1957) *A Method for Specification of Sand for Beach Fills*. Technical Memorandum 102, Beach Erosion Board, US Corps of Engineers.

Krumbein, W.C. (1963) *The Analysis of Observational Data from Natural Beaches*. Beach Erosion Board, Technical Memorandum 130.

Kunz, H. (1990) Artificial beach nourishment on Nordeney, a case study. *Proceedings 22nd Coastal Engineering Conference*, pp. 3254–3267.

Lankford, T.E. and Baca, B.J. (1989) Comparative environmental impacts of various forms of beach nourishment. *Proceedings Coastal Zone '89 Conference*, pp. 2046–2059.

Larson, M. and Kraus, N.C. (1991) Mathematical modeling of the fate of beach fill. *Journal of Coastal Engineering*, **16**: 83–114.

Larson, M., Kraus, N.C. and Byrnes, M.R. (1990) *SBEACH, Numerical Model for Simulating Storm Induced Beach Changes*. CERC Technical Report, 89–9.

Laubscher, W., Swart, D.H., Schoonees, D., Pfaff, W.M. and Davis, A.B. (1990) The Durban beach restoration scheme after 30 years. *Proceedings 22nd Coastal Engineering Conference*, pp. 3227–3238.

Laustrup, C. (1993) Coastal erosion management of sandy beaches in Denmark. *Proceedings of the Coastal Zone '93 Conference, New Orleans* (pp. not known).

Leadon, M.E. (1991) Littoral environmental considerations of a barrier island in

beach fill design: Key Biscayne. *Proceedings Coastal Sediments '91*, pp. 2089–2100.
Leatherman, S.P. (1986) Impacts of sea level rise on the coast of South America. In: J. Titus (ed.) *Impacts of Changes in Stratospheric Ozone and Global Climate*, vol. 4, pp. 73–82.
Leatherman, S.P. (1990) Modelling shore response to sea-level rise on sedimentary coasts. *Progress in Physical Geography*, **14**(4): 447–464.
Lelliott, R.E.L. (1989) The evolution of Bournemouth sea defences. In: *Coastal Management. Proceedings of the Conference organised by the Institution of Civil Engineering in Bournemouth*, 9–11 May 1989. Telford, London, pp. 263–277.
Leonard, L., Clayton, T. and Pilkey O.H. (1990a) An analysis of replenished beach design parameters on US East Coast barrier islands. *Journal of Coastal Research*, **6**: 15–36.
Leonard, L.A., Dixon, K.L. and Pilkey, O.H. (1990b) A comparison of beach replenishment on the US Atlantic, Pacific and Gulf coasts. *Journal of Coastal Research*, Special Issue, **6**: 127–140.
Linley-Adams, G. (1994) *The Good Beach Guide*. Readers Digest.
Lofty, M.F. and Frihy, O.E. (1993) Sediment balance in the nearshore zone of the Nile delta coast, Egypt. *Journal of Coastal Research*, **9**: 654–662.
Louisse, C.J. and Kuik, A.J. (1990) Coastal defence alternatives in The Netherlands. *Proceedings 22nd Conference on Coastal Engineering*, pp. 1.1–1.14.
Ly, C.K. (1980) The role of the Akosombo Dam on the Volta River in causing coastal erosion in central and eastern Ghana. *Marine Geology*, **37**: 323–332.
Macintosh, K.J. and Anglin, C.D. (1988) Artificial beach units on Lake Michigan. *Proceedings 21st Conference on Coastal Engineering*, pp. 2840–2854.
Manning-Saunders, R. (1951) *Seaside England*. Batsford, London.
Marsh, G.A and Turbeville, D.B. (1981) The environmental impact of beach nourishment: two studies in south-eastern Florida. *Shore and Beach*, **47**: 40–44.
Masselink, G. and Short, A.D. (1993) The effect of tide range on beach morphodynamics and morphology: a conceptual beach model. *Journal of Coastal Research*, **9**: 785–800.
Mather, R.S. and Ritchie, W. (1977) *The Beaches of the Highlands and Islands*. Countryside Commission for Scotland, Perth, pp. 94–98.
May, V.J. (1990) Replenishment of resort beaches at Bournemouth and Christchurch, England. *Journal of Coastal Research*, Special Issue, **6**: 11–16.
McFarland, S., Whitcombe, L. and Collins, M. (1994) Recent shingle beach renourishment schemes in the UK: some preliminary observations. *Ocean & Coastal Management*, **25**: 143–149.
McGrath, B.L. (1968) Erosion of Gold Coast beaches. *Journal of the Institute of Engineers, Australia*, **40**: 155–156.
McLachlan, A. and Erasmus, T. (eds.) (1983) *Sandy Beaches as Ecosystems*. Junk, The Hague.
McLean, R.F. (1978) Recent coastal progradation in New Zealand. In: J.L. Davies and M.A.J. Williams (eds.) *Landscape Evolution in Australasia*, Australian National University, Canberra, pp. 168–196.
McLennan, R.L. (1976) Snowy River Delta. In: J.G. Douglas and J.A. Ferguson (eds.) *Geology of Victoria*. Geology Society of Australia Special Publication, **5**: 325–327.
Moller, J.P. and Swart, D.H. (1987) Extreme erosion events on an artificially nourished beach. *Proceedings Coastal Sediments '87*, pp. 1882–1896.
Moller, J.P., Owen, K.C. and Swart, D.H. (1986) Coastal engineering studies for

inshore mining of diamonds at Oranjemund. *Proceedings 20th Coastal Engineering Conference*, pp. 1407–1418.

Møller, J.T. (1985) Denmark. In: E.C.F. Bird and M.L. Schwartz (eds.) *The World's Coastline.* Van Nostrand Reinhold, New York, pp. 325–333.

Møller, J.T. (1990) Artificial beach nourishment on the Danish North Sea coast. *Journal of Coastal Research*, Special Issue, **6**: 1–10.

Molnia, B.F. (1977) Rapid shoreline erosion at Icy Bay, Alaska. *Proceedings 11th Offshore Technology Conference*, vol. 4, pp. 115–126.

Morton, R.A., Leach, M.P., Paine, J.G. and Cardoza, M.A. (1993) Monitoring beach changes using GPS surveying techniques. *Journal of Coastal Research*, **9**: 702–720.

Nakayama, J.P., Kataoka, S. and Obara, K. (1982) Construction of artificial beaches in Japan. *Civil Engineering in Japan*, **21**: 100–113.

Nelson, W.G. (1993) Beach restoration in the southeastern US: environmental effects and biological monitoring. *Ocean & Coastal Management*, **19**: 157–182.

Nersesian, G.K. (1977) Beach fill design and placement at Rockaway Beach, New York, using offshore ocean borrow sources. *Proceedings Coastal Sediments '77*, pp. 228–247.

Newman, D.E. (1976) Beach replenishment: sea defences and a review of the role of artificial beach replenishment, *Proceedings of the Institute of Civil Engineers*, **60**: 445–460.

Nicholls, R.J. (1990) Managing erosion problems on shingle beaches: British experience. *Proceedings 3rd European Workshop on Coastal Zones*, pp. 2.35–2.36.

Nicholls, R.J. and Webber, N.B. (1987) Coastal erosion in the eastern half of Christchurch Bay. In: M.G. Culshaw et al. (eds.), *Planning and Engineering Geology. Engineering Geology Special Publication.* No. 4. Geological Society, London, pp. 549–554.

Nordstrom, K.F. and Allen, J.R. (1980) Geomorphologically compatible solutions to beach erosion. *Zeitschrift für Geomorphologie*, Supplementband **34**: 142–154.

Nordstrom, K.F., Allen, J.R., Sherman, D.J. and Psuty, N.P. (1979) Management considerations for beach nourishment at Sandy Hook, New Jersey. *Coastal Engineering*, **2**: 215–236.

Nordstrom, K.F., Psuty, N. and Carter, R.W.G. (1990) *Coastal Dunes: Form and Process.* Wiley, Chichester.

Norris, R.M. (1964) Dams and beach-sand supply in Southern California. In: R.L. Miller (ed.) *Papers in Marine Geology.* Macmillan, New York, pp. 154–171.

Norrman, J.O. (1980) Coastal erosion and slope development in Surtsey Island, Iceland. *Zeitschrift für Geomorphologie*, Supplementband **34**: 20–38.

Noyce, T. (1993) *Beach and offshore sand modelling in South Australia.* GIS Section, Department of Housing and Urban Development, Adelaide.

Nunny, R.S. (1978) *A survey of the dispersal of colliery waste from Lynemouth Beach, Northumberland.* Technical Report 43, MAFF, Great Yarmouth.

Olsen, E.J. (1982) South Seas Plantation Beach improvement project. *Shore and Beach*, **50**: 6–10.

Olsen, E.J. and Bodge, K.R. (1991) The use of aragonite as an alternative source of beach fill in south-east Florida. *Proceedings Coastal Sediments '91*, pp. 2130–2144.

Olson, J.S. (1958) Lake Michigan dune development. *Journal of Geology*, **66**: 254–263, 345–351, 473–483.

Orlova, G. and Zenkovich, V.P. (1974) Erosion on the shores of the Nile delta. *Geoforum*, **18**: 68–72.

Orme, A.R. (1985) California. In: E.C.F. Bird and M.L. Schwartz (eds.) *The World's Coastline*. Van Nostrand Reinhold, New York, pp. 27–36.

Orviku, K., Bird, E.C.F. and Schwartz, M.L. (1995) The provenance of beaches on the Estonian islands of Hiiumaa, Saaremaa and Muhu. *Journal of Coastal Research*, **11**: 96–106.

Parks, J. (1989) Beachface dewatering: a new approach to beach stabilisation. *The Compass*, **66**: 65–72.

Parry, G.D. and Collett, L.C. (1985) *The Control of Seagrass in the Rosebud–Rye Region of Port Phillip Bay*. Coastal Unit Technical Report No. 2, Ministry for Planning and Environment, Victoria, Australia.

Paskoff, R. (1994) *Les Littoraux. Impact des aménagements sur leur évolution*. 2nd edn. Masson, Paris.

Paskoff, R. and Petiot, R. (1990) Coastal progradation as a by-product of human activity: an example from Chañaral Bay, Atacama Desert, Chile. *Journal of Coastal Research*, Special Issue, **6**: 91–102.

Pattearson, C. and Carter, T. (1989) Coastal management demonstrations. *9th Annual Conference Coastal & Ocean Engineering*, pp. 403–406.

Pattearson, C. and Pattearson, D.C. (1983) Gold Coast Longshore Transport. *6th Annual Conference Coastal & Ocean Engineering*, pp. 211–217.

Pearson, D.R. and Riggs, S.R. (1981) Relationship of surface sediments on the lower forebeach and nearshore shelf to beach nourishment at Wrightsville Beach, NC. *Shore and Beach*, **49**: 26–31.

Perlin, R.M. and Dean, R.G. (1985) Models of bathymetric response to structures. *Journal of Waterways, Port, Coastal and Ocean Engineering*, **111**: 153–170.

Pethick, J. (1984) *An Introduction to Coastal Geomorphology*. Arnold, London.

Piccazzo, M., Firpo, M., Corradi, N. and Campi, F. (1992) Coastline changes on beaches affected by defence works: a case study in Albissola (Western Liguria, Italy). *Bollettino di Oceanologica Teorica ed Applicata*, **10**: 197–210.

Pilkey, O.H. (1990) A time to look back at beach replenishment (Editorial). *Journal of Coastal Research*, **6**: iii–vii.

Pilkey, O.H. (1995) The fox guarding the hen house (Editorial). *Journal of Coastal Research*, **11**: iii–v.

Pilkey, O.H. and Clayton, T.D. (1987) Beach replenishment: the national solution? *Proceedings Coastal Zone '87 Conference*, New York, pp. 1408–1419.

Pilkey, O.H. and Clayton, T.D. (1989) Summary of beach replenishment experience on US East Coast barrier islands. *Journal of Coastal Research*, **5**: 147–159.

Pilkey, O.H. and Leonard, L.A. (1990) Reply to Houston discussion. *Journal of Coastal Research*, **6**: 1047–1057.

Pilkey, O.H., Young, R.S., Riggs, S.R., Smith, A.W.S., Wu, H. and Pilkey, W.D. (1993) The concept of shoreface profile of equilibrium: a critical review. *Journal of Coastal Research*, **9**: 255–278.

Pirazzoli, P.A. (1986) Secular trends of relative sea-level changes indicated by tide-gauge records. *Journal of Coastal Research*, Special Issue, **1**: 1–26.

Pirazzoli, P.A. (1991) *World Atlas of Holocene Sea-level Changes*. Elsevier Oceanography Series, vol. 58.

Pluijm, M. (1990) Seaward coastal defence for the Dutch coast. *Proceedings 22nd Conference on Coastal Engineering*, pp. 2010–2017.

Pompe, J.J. and Rinehart, J.R. (1994) Estimating the effect of wider beaches on coastal housing prices. *Ocean & Coastal Management*, **22**: 141–152.

Pope, J. (1986) Segmented offshore breakwaters: an alternative for beach erosion control. *Shore and Beach*, **54**: 3–6.

Port Phillip Authority (1977) *Port Phillip Coastal Study*. Melbourne, Australia.

Psuty, N.P. (1984) USA New Jersey and New York. In: H.J. Walker (ed.) *Artificial Structures and Shorelines*. Louisiana State University, pp. 71–78.

Psuty, N.P. (1985) Surinam. In: E.C.F. Bird and M.L. Schwartz (eds.) *The World's Coastline*. Van Nostrand Reinhold, New York, pp. 99–101.

Psuty, N.P. and Moreira, M.E. (1990) Nourishment of a cliffed coastline, Praia da Rocha, the Algarve, Portugal. *Journal of Coastal Research*, Special Issue, **6**: 21–32.

Pugh, K.B., Andrews, A.R., Gibbs, C.F., Davis, S.J. and Floodgate, G.D. (1974) Some physical, chemical and microbiological characteristics of two beaches of Anglesey. *Journal of Experimental Marine Biology and Ecology*, **15**: 305–333.

Pullen, E.J. and Naqui, S.M. (1983) Biological impacts on beach replenishment and borrowing. *Shore and Beach*, **51**(2): 27–31.

Quigley, R.M. and De Nardo, L.R. (1980) Cyclic instability modes of eroding clay bluffs. *Zeitschrift für Geomorphologie*, Supplementband **34**: 39–47.

Ranwell, D.S. and Boar, R. (1986) *Coast Dune Management Guide*. Institute of Terrestrial Ecology, Huntingdon, UK.

Reilly, F.J. and Bellis, V.J. (1983) *The Ecological Impact of Beach Nourishment with Dredged Materials on the Intertidal Zone at Bogue Banks, North Carolina*. US Army Corps of Engineers, CERC, Fort Belvoir, Virginia. Miscellaneous Paper 83-3.

Riddell, K.J. and Young, S.W. (1992) The management and creation of beaches for coastal defence. *Journal of International Water Resources Management*, **139**: 588–597.

Robinson, A.H.W. (1955) The harbour entrances of Poole, Christchurch and Pagham. *Geographical Journal*, **121**: 33–50.

Robinson A.H.W. (1961) The hydrography of Start Bay and its relationship to beach change at Hallsands. *Geographical Journal*, **127**: 64–77.

Robinson, A.H.W. (1966) Residual currents in relation to shoreline evolution on the East Anglian coast. *East Midland Geographer*, **3**: 307–321.

Robinson, D.A. (1993) Tweed Heads sand by-passing scheme. *Bruun Workshop on Beach Nourishment and Sand By-passing*. Department of Civil Engineering, University of Queensland, pp. 28–34.

Roellig, D.A. (1989) Shoreline response to beach nourishment. *Proceedings Coastal Zone '89*, pp. 2104–2109.

Roelse, P. (1990) Beach and dune nourishment in the Netherlands. *Proceedings 22nd International Engineering Conference*, pp. 1984–1997.

Roelvink, J.A. (1989) Feasibility of offshore nourishment of Dutch sandy coast. *International Conference on Hydraulic and Environmental Modelling of Coastal, Estuarine and River Waters, Bradford*, pp. 107–114.

Rotnicki, K. (ed.) (1994) *Changes of the Polish Coastal Zone*. Quaternary Research Institute, Adam Mickiewicz University, Poznan.

Rouch, F. and Bellessort, B. (1990) Man-made beaches more than 20 years on. *Proceedings 22nd Coastal Engineering Conference*, pp. 2394–2401.

Royal Commission on Coastal Erosion (1907–11) *Reports* (3 volumes). HMSO, London.

Sanlaville, P. (1985) Arabian Gulf coasts. In: E.C.F. Bird and M.L. Schwartz (eds.) *The World's Coastline*. Van Nostrand Reinhold, New York, pp. 729–733.

Schou, A. (1952) Direction determining influence of the wind on shoreline simplification and coastal dunes. *Proceedings 17th Congress International Geographical Union*, Washington, DC, pp. 370–373.

Schwartz, M.L. (1967) The Bruun theory of sea level rise as a cause of shore erosion. *Journal of Geology*, **75**: 76–92.

Schwartz, M.L. (1982) *The Encyclopedia of Beaches and Coastal Environments*. Hutchinson and Ross, Stroudsburg, PA.
Schwartz, M.L. and Bird, E.C.F. (eds.) (1990) Artificial beaches. *Journal of Coastal Research*, Special Issue, **6**.
Schwartz, M.L. and Terich, T.A. (1985) Washington. In: E.C.F. Bird and M.L. Schwartz (eds.) *The World's Coastline*. Van Nostrand Reinhold, New York, pp. 17–22.
Schwartz, M.L., Mahala, J. and Bronson, H.S. (1985) Net shore-drift along the Pacific coast of Washington State. *Shore and Beach*, **53**: 21–25.
Schwartz, M.L., Marti, J.L.J., Herrera, J.F. and Montero, G.G. (1991) Artificial nourishment at Varadero Beach, Cuba. *Coastal Sediments '91*, pp. 2081–2088.
Schwartz, R.K. and Musialawski, F.R. (1977) Nearshore disposal: onshore sediment transport. *Proceedings Coastal Sediments '77*, pp. 85–101.
S.C.O.R Working Group 89 (1991) The response of beaches to sea-level changes: a review of predictive models. *Journal of Coastal Research*, **7**(3): 895–921.
Shave, K.G. (1989) Beach management. In: *Coastal Management. Proceedings of the Conference organised by the Institution of Civil Engineering in Bournemouth*, 9–11 May 1989. Telford, London, pp. 177–186.
Shepard, F.P. and Wanless, H.R. (1971) *Our Changing Coastlines*. McGraw-Hill, New York.
Sherlock, R.L. (1922) *Man as a Geological Agent. An Account of his Action on Inanimate Nature*. Witherby, London.
Short, A.D. (1992) Beach systems of the central Netherlands coast: processes, morphology and structural impacts in a storm driven multi-bar system. *Marine Geology*, **107**: 103–137.
Short, A.D. (1996) *Beach Morphodynamics*. Wiley, Chichester, in press.
Short, A.D. and Hesp, P.A. (1982) Wave, beach and dune interactions in southeastern Australia. *Marine Geology*, **48**: 259–284.
Short, A.D. and Hogan, C.L. (1994) Rip current and beach hazards: their impact on public safety and implications for coastal management, *Journal of Coastal Research*, Special Issue, **12**: 197–209.
Short, A.D., Williamson, B. and Hogan, C.L. (1993) The Australian Beach Safety and Management Progam – Surf Life Saving. Australia's approach to beach safety and coastal planning. *Proceedings 11th Australasian Conference on Coastal and Ocean Engineering*, pp. 113–118.
Shuisky, Y.D. (1994) An experience of studying artificial ground terraces as a means of coastal protection. *Ocean & Coastal Management*, **22**: 127–139.
Shuisky, Y.D. and Schwartz, M.L. (1979) Natural laws in the development of artificial sandy beaches. *Shore and Beach*, **47**: 33–36.
Silvester, R. (1970) Growth of crenulate shaped bays to equilibrium. *Journal of the Waterways and Harbors Division, Proceedings American Society of Civil Engineers*, **96**: 275–287.
Silvester, R. (1974) *Coastal Engineering*. Elsevier, Amsterdam.
Silvester, R. (1976) Headland defense of coasts. *Coastal Engineering*, **1**: 1394–1406.
So, C.L. (1974) Some coast changes around Aberystwyth and Tanybwlch, Wales. *Transactions Institute of British Geographers*, **62**: 143–153.
So, C.L. (1982) Wind-induced movements of beach sand at Portsea, Victoria. *Proceedings of the Royal Society of Victoria*, **94**: 61–68.
Spaulsing, I.A. (1973) *Factors Related to Beach Use*. University of Rhode Island, Marine Technical Report Series.
Steers, J.A. (1953) *The Sea Coast*. Collins, London.

Steers, J.A. (1964) *The Coastline of England and Wales*. Cambridge University Press, Cambridge.

Stive, M.J.F., Nicholls, R.L. and De Vriend, H.J. (1991) Sea level rise and shore nourishment: a discussion. *Journal of Coastal Engineering*, **16**: 147–163.

Stronge, W. (1990) Economic and fiscal impact of beach restoration: the case of Captivia. *Proceedings 35th Annual Meeting of the Florida Shore and Beach Preservation Association*, Tallahassee, pp. 74–86.

Sudar, R.A., Pope, J., Hillyer, T. and Crumm, J. (1995) Shore protection projects of the US Army Corps of Engineers. *Shore & Beach*, **63**: 3–16.

Sunamura, T. (1992) *Geomorphology of Rocky Coasts*. Wiley, Chichester.

Tait, J.F. and Griggs, G.B. (1990) Beach response to the presence of a sea wall: a comparison of field observations. *Shore and Beach*, **58**: 11–28.

Tanner, W.F. (1987) The beach: where is the 'river of sand'?. *Journal of Coastal Research*, **3**: 377–386.

Tanner, W.F. and Stapor, F. (1972) Accelerating crisis in beach erosion. *International Geography*, **2**: 1020–1021.

Teh Tiong Sa (1985) Peninsular Malaysia. In: E.C.F. Bird and M.L. Schwartz (eds.) *The World's Coastline*. Van Nostrand Reinhold, New York, pp. 789–795.

Terchunion, A.V. (1989) Performance of the Stabeach System on beach preservation technology at Hutchinson Island, Florida. *Proceedings Florida Shore and Beach Preservation Association*, February 1989, pp. 229–238.

Terich, T.A. and Komar, P. D. (1974) Bayocean spit, Oregon: history of development and erosional destruction. *Shore and Beach*, **42**: 3–10.

Thompson, C.L. (1987) Beach nourishment monitoring on the Great Lakes. *Proceedings Coastal Sediments '87*, pp. 1203–1215.

Thorn, R.B. (1960) *The Design of Sea Defence Works*. Butterworth, London.

Thyme, F. (1990) Beach nourishment on the West Coast of Jutland. *Journal of Coastal Research*, **6**: 201–209.

Titus, J.G. (1986) The causes and effects of sea level rise. In: J.G. Titus (ed.) *Effects of Changes in Stratospheric Ozone and Global Climate. 1. Overview*. US Environment Protection Agency, pp. 219–249.

Tower, J. and Kain, A. (1993) *Victorian Coastal Recreation Study*. Department of Sport and Recreation, Victoria, Australia.

Townend, I.H. and Fleming, C.A. (1991) Beach nourishment and socio-economic aspects. *Journal of Coastal Engineering*, **16**: 115–127.

Turner, I.L. (1995) Simulating the influence of groundwater seepage on sediment transported by the sweep of the swash zone across macro-tidal beaches. *Marine Geology*, **125**: 153–174.

US Army Corps of Engineers (1984) *Shore Protection Manual* (2 volumes). 4th edn.

US Army Corps of Engineers (1994) *Shoreline Protection and Beach Erosion Control Study. Phase 1. Cost Comparison of Shoreline Protection Projects of the US Army Corps of Engineers*.

Usoro, E.J. (1985) Nigeria. In: E.C.F. Bird and M.L. Schwartz (eds.) *The World's Coastline*. Van Nostrand Reinhold, New York, pp. 607–613.

Vallianos, L. (1974) Beach fill planning – Brunswick County, North Carolina. *Proceedings 14th Coastal Engineering Conference*, pp. 1350–1369.

Van de Graaff, J., Niemeyer, H.D. and Van Overveen, J. (1991) Beach nourishment philosophy and coastal protection policy. *Journal of Coastal Engineering*, **16**: 3–22.

Vera Cruz, D. (1972) Artificial nourishment at Copacabana Beach. *Proceedings 13th Coastal Engineering Conference*, pp. 1451–1463.

Vermeer, D.E. (1985) Mauritania. In: E.C.F. Bird and M.L. Schwartz (eds.) *The World's Coastline*. Van Nostrand Reinhold, New York, pp. 549–553.

Vesterby, H. (1988) Beach management with the 'coastal drain system'. Paper presented to the *Beach Preservation Technology Conference*, Gainesville, FL.

Viles, H. and Spencer, T. (1995) *Coastal Problems: Geomorphology, Ecology and Society at the Coast*. Arnold, London.

Wace, N. (1994) Beachcombing for ocean litter. *Australian Natural History*, **23**: 46–52.

Wace, N. and Zann, L.P. (1995) *Ocean and Beach Litter. State of the Marine Environment, Australia*. Technical Paper. Great Barrier Reef Marine Park Authority, Townsville.

Walker, H.J. and Mossa, J. (1986) Human modification of the shoreline of Japan. *Physical Geography*, **7**: 116–139.

Walton, T.L. and Dean, R.G. (1976) Use of outer bars of inlets as sources of beach nourishment material. *Shore and Beach*, **44**: 13–19.

Walton, T.L. and Purpura, J.S. (1977) Beach nourishment along the southeast Atlantic and Gulf coasts. *Shore and Beach*, **45**: 10–18.

Watson, J.E. (1973) The impact of the ethane pipeline on the marine ecosystems of Port Phillip Bay, Victoria. *Victorian Naturalist*, **90**: 60–65.

Weggel, J.R. (1986) Economics of beach nourishment under scenario of rising sea level. *Journal of Waterways, Port, Coastal and Ocean Engineering*, **112**: 418–427.

Weggel, J.R. and Sorensen, R.M. (1991) Performance of the 1986 Atlantic City, New Jersey, beach nourishment project. *Shore and Beach*, **55**: 2–11.

West, N.C. and Heatwole, C.A. (1979) Urban beach use: ethnic background and socio-environmental attitudes. *Proceedings 5th Annual Conference Coastal Society*, pp. 195–204.

Wiegel, R.L. (1987) Trends in coastal engineering management. *Shore and Beach*, **55**: 2–11.

Wigley, T.M.L. and Raper, S.C.A. (1993) Future changes in global mean temperature and sea level. In: R.A. Warrick, E.M. Barrow and T.M.L. Wigley (eds.) *Climate and Sea Level Change*. Cambridge University Press, Cambridge, pp. 111–133.

Willis, D.H. and Price, W.A. (1975) Trends in the application of research to solve coastal engineering problems. In: J.R. Hails and A.P. Carr (eds) *Nearshore Sediment Dynamics and Sedimentation*, Wiley, New York, pp. 111–122.

Willmington, R.H. (1983) The renourishment of Bournemouth beach 1974–1975. *Proceedings of Shoreline Protection Conference*, 14–15 September 1982, Telford, London, pp. 157–162.

Wong, P.P. (1985) Artificial coastlines: the example of Singapore. *Zeitschrift für Geomorphologie*, Supplementband **57**: 175–192.

Wong, P.P. (1991) Seaside resorts in Singapore: a geomorphological perspective. In: L.S. Chia and L.M. Chou (eds.) *Urban coastal area management: the experience of Singapore*. ICLARM Conference Proceedings, vol. 25, pp. 109–116.

Woodworth, P. (1991) The Permanent Service for Mean Sea Level and the Global Sea Level Observing System. *Journal of Coastal Research*, **7**: 699–710.

Wright, L.D. and Short, A.D. (1984) Morphodynamic variability of surf zones and beaches: a synthesis. *Marine Geology*, **56**: 93–118.

Wynne, A.A. (1984) *Adelaide Coast Protection Strategy Review*. Coast Protection Board, South Australia.

Yasso, W.E. (1965) Fluorescent tracer particle determination of the size velocity relation for foreshore sediment transport. *Journal of Sedimentary Petrology*, **35**: 989–993.

Zenkovich, V.P. (1967) *Processes of Coastal Development.* Translated by O.D. Fry, J.A. Steers and C.A.M. King (eds.). Oliver and Boyd, Edinburgh.

Zenkovich, V.P. (1973) Geomorphological problems of protecting the Caucasian Black Sea coast. *Geographical Journal,* **139**: 460–466.

Zenkovich, V.P. and Schwartz, M.L. (1987) Protecting the Black Sea–Georgian SSR gravel coast. *Journal of Coastal Research,* **3**: 201–209.

Zunica, M. (1971) *La spiagge del Venete*. Presso Istituo de Geografia, Università di Padova.

Zunica, M. (1976) Coastal changes in Italy during the past century. *Italian Contributions, 23rd International Geographical Congress,* pp. 275–281.

Zunica, M. (1990) Beach behaviour and defences along the Lido di Jesolo, Gulf of Venice, Italy. *Journal of Coastal Research,* **6**: 709–719.

Author Index

Geographical Index

Subject Index